Geotechnologies and the Environment

Volume 20

Series editors
Jay D. Gatrell, *Department of Geology & Geography, Eastern Illinois University, Charleston, IL*
Ryan R. Jensen, *Department of Geography, Brigham Young University, Provo, UT, USA*

The *Geotechnologies and the Environment* series is intended to provide specialists in the geotechnologies and academics who utilize these technologies, with an opportunity to share novel approaches, present interesting (sometimes counter-intuitive) case studies, and, most importantly, to situate GIS, remote sensing, GPS, the internet, new technologies, and methodological advances in a real world context. In doing so, the books in the series will be inherently applied and reflect the rich variety of research performed by geographers and allied professionals.

Beyond the applied nature of many of the papers and individual contributions, the series interrogates the dynamic relationship between nature and society. For this reason, many contributors focus on human-environment interactions. The series is not limited to an interpretation of the environment as nature per se. Rather, the series "places" people and social forces in context and thus explores the many socio-spatial environments humans construct for themselves as they settle the landscape. Consequently, contributions will use geotechnologies to examine both urban and rural landscapes.

More information about this series at http://www.springer.com/series/8088

Stavros Kolios • Andrei V. Vorobev
Gulnara R. Vorobeva • Chrysostomos Stylios

GIS and Environmental Monitoring

Applications in the Marine, Atmospheric and Geomagnetic Fields

Springer

Stavros Kolios
Laboratory of Knowledge and Intelligent Computing, Department of Computer Engineering
Technological Educational Institute of Epirus
Kostakioi, Arta, Greece

Gulnara R. Vorobeva
Computer Science and Robotics Department
Ufa State Aviation Technical University
Ufa, Russia

Andrei V. Vorobev
Computer Science and Robotics Department
Ufa State Aviation Technical University
Ufa, Russia

Chrysostomos Stylios
Laboratory of Knowledge and Intelligent Computing, Department of Computer Engineering
Technological Educational Institute of Epirus
Kostakioi, Arta, Greece

ISSN 2365-0575 ISSN 2365-0583 (electronic)
Geotechnologies and the Environment
ISBN 978-3-319-53084-0 ISBN 978-3-319-53086-4 (eBook)
DOI 10.1007/978-3-319-53086-4

Library of Congress Control Number: 2017943535

Printed on acid-free paper

This Springer imprint is published by Springer Nature
The registered company is Springer International Publishing AG
The registered company address is: Gewerbestrasse 11, 6330 Cham, Switzerland

Preface

This book constitutes a notable contribution to investigate and present the capabilities of Geographic Information Systems (GIS) and their applicability and usefulness in environmental-related applications and sciences. The focus is on the design, creation, development, and operation of integrated Web-based GIS applications for the weather, the marine and atmospheric environments, and the Earth's magnetic field.

The book provides an overview of the latest developments in the field of environmental GIS as well as novel approaches and methodologies to develop integrated GIS applications in the field of geosciences. More specifically, the aim of this book is to present characteristic applications of GIS to environmental monitoring including GIS solutions for eco-mapping sea and port-related parameters, climate changes, and geomagnetic field.

The first part of the book presents the state of the art of GIS tools and methods as well as an analytic description to develop integrated environmental GIS applications. For this scope, different innovative approaches, such as cloud GIS and Google Apps for GIS, are used, justifying the merit of WebGIS in the world of environmental applications.

The second part of the book provides an overview of geomagnetic field parameters and reveals the potential of using GIS for modeling and analyzing the Earth's magnetic (geomagnetic) field and its parameters. Here, the authors present the recently introduced phenomenon called the "geomagnetic pseudostorm", which is modeled and further analyzed with GIS technology and tools.

Due to the extensive description of the GIS applications, this book appeals to people interested in various areas where spatial information is of paramount relevance (e.g., social and economic research and mapping, environmental and climate research, decision support systems, public services, and especially for geomagnetic field variations and for the design of warning systems for natural disasters). It presents modern methods and approaches to visualize and analyze spatial information using innovative techniques, procedures, and tools of WebGIS technology. In this book, readers will find a valuable companion in their efforts to design and develop their own WebGIS applications, as it includes useful examples of developing (Web)GIS applications regarding the monitoring of marine and atmospheric environments, as well as applications that deal with meteorological issues and the Earth's magnetic field along with solar activity (space weather information). This book can also serve as a useful reference source for graduates, researchers, and professionals related to the areas indicated above.

The book consists of two parts including two chapters each. The first part is dedicated to GIS basics and applications in the marine and atmospheric environments, while the second part is devoted to the theoretical background of the geomagnetic field and related GIS applications.

In the first part of the book, Chap. 1 is an introductory chapter that offers to readers a broad perspective of the fundamental principles and methods in spatial analysis along with definitions, an overview, and tools of GIS. Capabilities and procedures of GIS in data management and mapping of multilayered datasets are provided, accompanied by many explanatory figures to better explain and illustrate the visualization power of the approach. The role of WebGIS is stressed, and the main definitions and types of Web mapping and geospatial services are presented. Chapter 1 briefly describes the main software programming interfaces and modern software approaches and architectures for developing WebGIS, and includes all the required notation and terminology used throughout the book. This chapter is useful for any reader interested in learning about the main introductory issues regarding GIS and WebGIS development.

Chapter 2 is dedicated to the presentation of all the appropriate stages for the development of modern Web-based GIS application examples concerning marine and atmospheric environments. The methods, tools, and procedures used to develop four applications are analytically described. More specifically, a WebGIS platform for marine monitoring and visualizing key environmental aspects in port areas is described, and the used data and the user interface of the platform are presented. Then, a second WebGIS application focusing on monitoring and forecasting weather conditions near port areas in the Ionian and Adriatic Seas is presented. This application is a useful tool that provides not only marine weather forecasts to support mercantile and passenger traffic but also information concerning past weather events and the climate profile of the pilot port area. Moreover, this application operates as an alert system, providing notifications for extreme weather conditions. The third application refers to the development of a WebGIS application that provides spatial information about the locations in Greece where extreme weather events (more precisely regarding heavy precipitation, floods, hail, snowfall, and strong winds) occurred in a 20-year period (1992–2012). This application constitutes a characteristic example of a cloud-based GIS. The fourth example is an experimental WebGIS application for cloud storm monitoring/forecasting. This application practically visualizes, in a modern Web-based graphical user interface, all the final data products automatically collected from a prototype statistical model for storm monitoring/forecasting. The initial data came from meteorological satellites. This application provides valuable information about storms and their results (precipitation, rain, and lightning) over the great Mediterranean Basin.

Conclusively, Chap. 2 contains valuable information and characteristic examples and approaches useful for readers interested in developing their own WebGIS/cloud GIS applications.

The second part of the book is dedicated to the development and use of Web-based GIS applications as valuable tools serving experts in their effort to monitor and study an intriguing and—not rarely—hazard-generating natural phenomenon, geomagnetic activity. Although geomagnetic variations are not so "popular" in scientific circles, and thus are not as adequately studied as other phenomena in the atmosphere and hydrosphere, their description and monitoring are of great importance because of the negative influence of the Earth's magnetic field on a wide range of biological organisms and technical instrumentation. Moreover, the spatiotemporal variations of the geomagnetic field are extended over a wide range of time and spatial scales depending on the location, the season, the time period, and the interaction with solar wind. This spatial character of the geomagnetic data requires the application of GIS and related technologies to best process and interpret them.

In this context, Chap. 3 describes the mathematical fundamentals of geomagnetic field as a necessary basis for implementing methods and technologies of spatial analysis to model, monitor, and interpret the set of parameters of the Earth's magnetic field and its variations. The main definitions and classification of geomagnetic variations are presented, and some modeling and calculation methods for geomagnetic field are analytically described, as well. To outline the importance of geomagnetic field study, some examples of the influence of geomagnetic variations on biological and technical objects and systems are given. Moreover, the main methods and algorithms for detailed analysis of geomagnetic field parameters, including geomagnetic data analysis in both time and frequency domains, are described. Chapter 3 is useful for any reader interested in mathematical basics and algorithms for the development of applications for monitoring and modeling of the geomagnetic field and its variations.

Chapter 4 is dedicated to the presentation of applying modern Web-based GIS applications suitable for visualizing and interpreting the geomagnetic environment. Most attention is paid to the geoinformation Web portal called GEOMAGNET, which utilizes the capabilities provided by GIS in synergy with related information technologies. This portal, available at http://www.geomagnet.ru, monitors and analyzes parameters related to the geomagnetic field and its variations. It provides useful tools and a variety of services that are used by specialists in geophysics, geology, and other scientific fields. More specifically, the Geomagnetic Calculator (WMM2015) service provides and calculates the parameters of the normal geomagnetic field at a specific point with given geographical coordinates, which are obtained with the GIS support of the portal (geolocation, interactive mapping, reverse geocoding, etc.). The second service, called Solar Activity, integrates tools to monitor space weather parameters and visualizes the results of their latest 3-day monitoring and analysis. Moreover, this service operates as an alert system and provides notifications about geomagnetic storms. The third service, called Magnetic Stations, provides current geomagnetic information from ground-based instrumentation on the Earth's surface. The data are spatiotemporal and are collected from specific ground stations all around the world. The spatial nature of the magnetic stations' data requires GIS visualizing tools to interpret them. Conclusively, Chap. 4 contains valuable information, approaches, and examples of geomagnetic field 2D/3D visualization, which can be useful for readers interested in the study of geomagnetic data distribution on the Earth's surface.

The Greek authors would like to take this opportunity to acknowledge the partial support received from the European Union, as the applications presented in Chap. 2 of this book were partially developed within two projects: the *pAssengeRs and loGistics information Exchange System* (ARGES-I52.2) project financed under the European Territorial Cooperation Programme Greece–Italy 2007–2013, cofunded by the European Union European Regional Development Fund (ERDF) and by national funds of Greece and Italy, and the *Transnational ENhancement of ECOPORT8 network* (TEN ECOPORT) project, with code SEE/D/0189/2.2/X cofinanced by the European Union within the South East Europe Transnational Cooperation Programme and by national funds of Greece.

The Greek authors would like to express their thanks to colleagues, collaborators, and friends, namely L. Damiani, M. Mali, T. Branca, T. Floqi, J. Marinski, V. Hadzhiev, P. Petkov, M. Mega, L. Traversa, K. Grinias, P. Karamesini, S. Antipas, A. Cheildaris, and G. Mannarini, for their collaboration in the framework of the abovementioned projects. We would also like to express our gratitude to graduate students and Laboratory of Knowledge and Intelligent Computing staff, S. Petsios, K. Tsalikidis, P. Kokkinis, and D. Loukadakis, for their technical support and assistance during the development of the applications.

Finally, the authors would like to express their utmost gratitude to Ms. Andriani Oikonomou for her dedication and hard work in providing truly professional assistance in book editing, organizing all the final material, and supporting us in the preparation of this book with her continuous encouragement.

Kostakioi, Arta, Greece Stavros Kolios
Ufa, Russia Andrei V. Vorobev
Ufa, Russia Gulnara R. Vorobeva
Kostakioi, Arta, Greece Chrysostomos Stylios

Contents

Part I
GIS Basics and Applications in the Marine and Atmospheric Environments

Chapter 1
Geographic Information Systems

1.1 Brief History

The concept of geographic information systems (GIS) was introduced to cover all the essential needs of the scientific community to provide spatial information, analyze data, and create digital thematic maps through a computer. A fundamental for GIS is the overlaying of different kinds of information and data. Professor Ian L. McHarg (1920–2001) was one of the first scientists who described the GIS concept and the usefulness of map overlaying. In 1954, he took a position as Professor of Landscape Architecture and Regional Planning at the University of Pennsylvania and during the mid-1960s he came to be recognized worldwide for introducing ecological concerns into land planning (McHarg 1992). He introduced the concept of overlaying (Fig. 1.1) and highlighted the four basic principles (*measuring*, *mapping*, *monitoring*, and *modeling*) that founded GIS.

Nevertheless, the first documentation of what could be considered as the first GIS-like application was in France in 1832. More specifically, a French geographer, named Charles Picquet, created a map-based representation of cholera epidemiology in Paris (http://gallica.bnf.fr/ark:/12148/bpt6k842918) by identifying the 48 districts of Paris with different colors (Seine 1971). Following this concept, in 1854, John Snow (Johnson 2006) represented cholera deaths in London using points on a map (Fig. 1.2). The next significant step in the development of modern GIS was a printing technique known as *photozincography* proposed by Sir Henry James (1803–1877) in the mid-nineteenth century, which was used to separate out a layer of information from a map (Mumford 1999).

During the decade of the 1960s, the evolution of computer technology created a fruitful background for more organized efforts—from various scientific groups and academic communities—made toward the development of applications and procedures for multilayered data representation. One of the first and most characteristic integrated efforts was an operational application launched in Ottawa, Canada, which was used to collect, store, and analyze data about land use/land cover

S. Kolios et al., *GIS and Environmental Monitoring*, Geotechnologies and the Environment 20, DOI 10.1007/978-3-319-53086-4_1

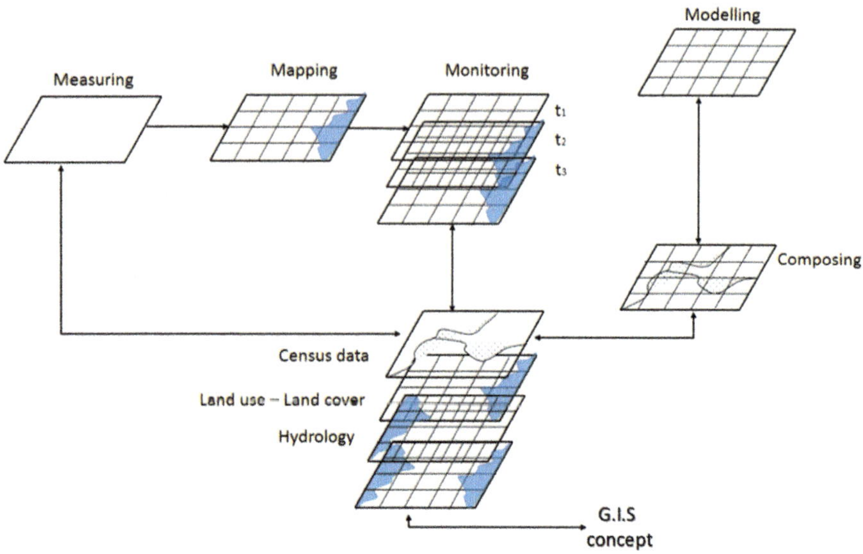

Fig. 1.1 The "four-M" concept according to McHarg's ideas in the creation of digital multicharacteristic map representations (Reproduced from: http://web.mst.edu/~rogersda/gis/History%20of%20GIS.pdf)

(LULC) in Canada (Tomlinson 1967). For this reason, Tomlinson is considered the "father" of modern GIS.

In the decades that followed, GIS software and related applications gradually evolved taking advantage of the advances in computer technology. As a result, GIS applications gradually have started to develop and operate in various businesses and in the defense sector, as well as in governmental and/or nongovernmental organizations worldwide. More specifically, the rapid evolution of computer technology, the spread of Internet access, the improvement of Internet connections, and the role of satellites for GIS software and the related applications bloomed. Key end users are governmental/nongovernmental bodies, municipalities, the private sector, and research and academic organizations interested in spatial information representation and analysis. These bodies have started to create digital maps and to analyze and share spatial data and information at a continuously lower cost as this technology became more and more easily accessible—from technical and economic points of view—to the authorities and the public.

Nowadays, many different GIS applications, using modern computers and Web technologies and capabilities, have become part of our daily life through personal computers and mobile devices (Fig. 1.3). GIS-related applications are being used by almost all scientific communities, the private and public sectors, and the public, whenever and wherever there is a need for spatial information and visualization.

Fig. 1.2 Region of Soho (London) map depicting a cholera outbreak (1855) (Reproduced from http://www.york.ac.uk/depts/maths/histstat/snow_map.htm)

1.2 GIS Definition and Overview

What exactly can be considered a geographic information system (GIS)? A GIS is an integrated system designed to create, capture, store, analyze, manage, and visualize all types of spatial or geographical data and information (e.g., Goodchild 1985; Burrough 1986). Such systems are considered to be essential tools—among others—for the scientific field of geoinformatics. GIS uses fundamental principles of geography, cartography, and geodesy in order to allow end users to create queries, analyze spatial information, provide data in maps, and present the final results of all these operations through detailed thematic digital maps (e.g., Clarke 1986; Maliene et al. 2011).

Through such systems, spatial data and information can be retrieved and integrated, and automated applications (static or dynamic) can be developed to cover an

Fig. 1.3 Adaptability of GIS applications to any computer or mobile device

extensive range of fields of interest, such as engineering, planning, management, transport/logistics, insurance, telecommunications, and the environment. Many GIS applications have been increasingly developed in—but are not limited to—the sectors of:

- Defense
- Governmental administration
- Construction
- Geophysical research
- Environmental protection/natural disaster management
- Atmospheric sciences
- Agriculture and forestry
- Commerce
- Transportation
- Telecommunications
- Health/medical resource management
- Archaeology

The majority of the GIS applications may use shared initial datasets, but where necessary, an administrative authority can create its own spatial data and information for its own GIS. Most users are involved exclusively in viewing aspects regarding GIS data. Significantly smaller groups of users are involved in the spatial data analysis and/or are involved in initial/thematic/attribute data provision and updating.

The structural characteristic of GIS is the data overlay. More analytically, every kind of data is a different data layer. Through GIS, many different layers of data can be analyzed and combined, generating final data and products. The integration of different layers of data gives the capability for final thematic digital maps, responding to user needs, to be constructed (Fig. 1.4). What is only needed for

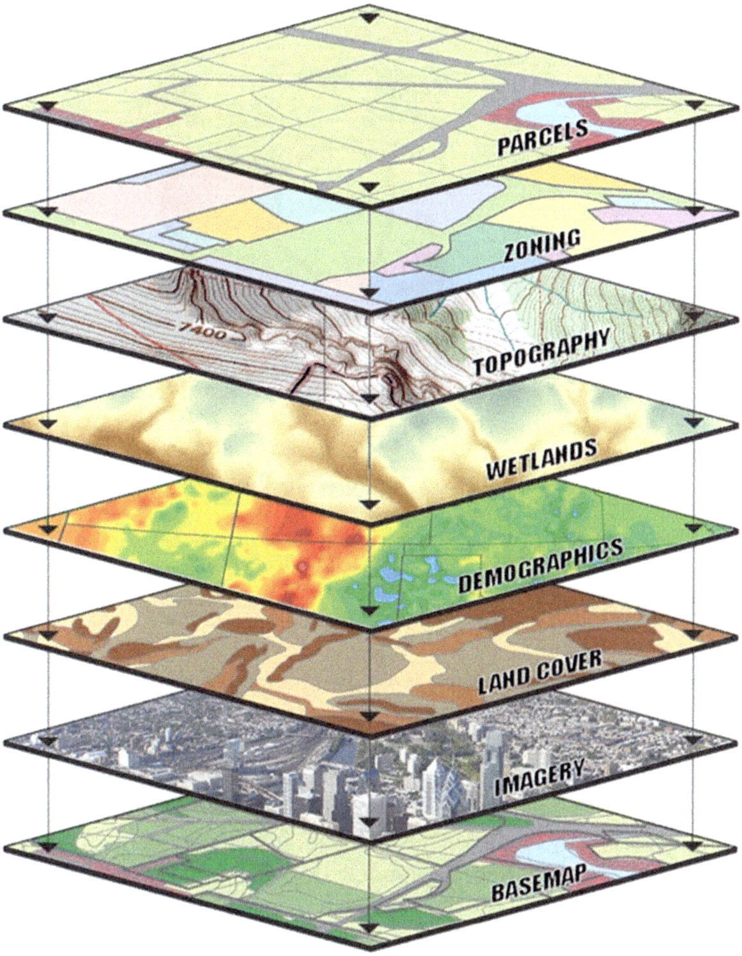

Fig. 1.4 Different layers of data can be combined through a GIS to represent realistic and integrated digital maps of the Earth's surface (Source: http://www.turfimage.com/)

the combination of different data layers are some common attributes, features, and the same datum and projection system in all the combined layers.

A GIS, as an integrated system, consists of five main parts (Fig. 1.5):

- Hardware (computers, servers, digitizers, scanners, and printers)
- Software (operating system, GIS application)
- Users (basic users with different privileges who provide service, analyze data, organize procedures, and define final products, as well as end users who view/obtain final data and products)
- Supported types of data (any kind of spatial data of vector and raster types, as well as attribute data)
- Procedures (e.g., input/capture data, data management, spatial analysis and modeling)

Fig. 1.5 GIS basic components

These general parts of the GIS are analyzed in Sects. 1.3, 1.4, 1.5, 1.6, and 1.7.

1.3 GIS Hardware

GIS software runs on a wide range of hardware types, from centralized computer servers to desktop computers used in stand-alone or networked configurations. Most GIS have 32-bit or 64-bit architecture, have a large main memory and storage capacity, use the UNIX operating system, and utilize a high-resolution graphics screen. They typically function as stand-alone systems, though they may be integrated into a network if required. Nevertheless, in many modern GIS there is the capability to operate as multiuser centralized systems (servers or host computers, which facilitate sharing between the various devices) or wide-area-network (WAN) systems where all GIS operations can have access to databases, which might reside almost anywhere. For personal use, almost all GIS can operate on most common laptops or desktop PCs which, nowadays, have multiple processors, at least 2 GB of RAM, an SVGA display, and a hard disk of many gigabytes.

1.4 GIS Software

GIS software provides the functions and tools needed to store, analyze, and display geographic information. Key software components of GIS software are considered to be:

- Tools and menus for the input/export and management of geographic information
- A database management system (DBMS) to store/organize geographic information
- Tools that support geographic queries, spatial analysis, and visualization
- A graphical user interface (GUI) for easy access to tools by end users

It should also be mentioned that the software should also support a wide range of peripherals by having the relevant drivers, i.e., the routines that handle the specific details and characteristics of a single peripheral device.

1.5 GIS Users

GIS technology is of limited value without people to manage the system and develop plans for application to real-world problems. GIS users range from technical experts who design and maintain such systems to those who use GIS to help them perform their daily work.

1.6 Supported Types of Data

A GIS is able to receive and analyze all types of data, which can be divided into two general categories: the *vector data type* and the *raster data type*.

In the *vector data type*, the data are stored as points, lines, and polygons. Less computer memory is used and better position accuracy is provided in relation to the raster format. The vector data type is useful for storing data that have discrete spatial boundaries, such as country borders, land parcels, and streets. The vector data type records and displays coordinates of objects with complete measurement accuracy in respect to ground measurements. The vector data type contains a lesser volume of information in relation to the raster data type for the same area. Additionally, it is easy to apply alphanumeric attributes to the defined schemes that express physical objects with points, lines, or polygons (Fig. 1.6). The calculation of vector positions and intersections follows the principles of analytic geometry.

In the *raster data type* (Fig. 1.7b, d), data are represented by pixels with values, creating a grid, allowing certain types of operations that could not be performed with the *vector data type* (Fig. 1.7a, c). Using the raster data type, Map Algebra is used along with multiple data layers to create index maps. The raster data type is useful for storing data that vary continuously, as in the case of aerial photographs, satellite images, surfaces of chemical concentrations, and/or elevation surfaces. Raster data consist of a regular two-dimensional (2D) grid of cells (pixels). The grid is characterized by a georeferenced origin, its georeferenced orientation, and the raster cell size (pixel size). Raster data may also be arranged in three dimensions. In this case, the three-dimensional (3D) cell becomes a cube (a voxel). The geometric accuracy of raster data is limited by the pixel resolution. In raster data, apart from geometric correction procedures, radiometric transformations can be applied. Furthermore, Boolean algebra operators can be used, allowing data combination among different raster layers.

With the combination of the raster and vector data types, a realistic representation of the world can be achieved (Fig. 1.8).

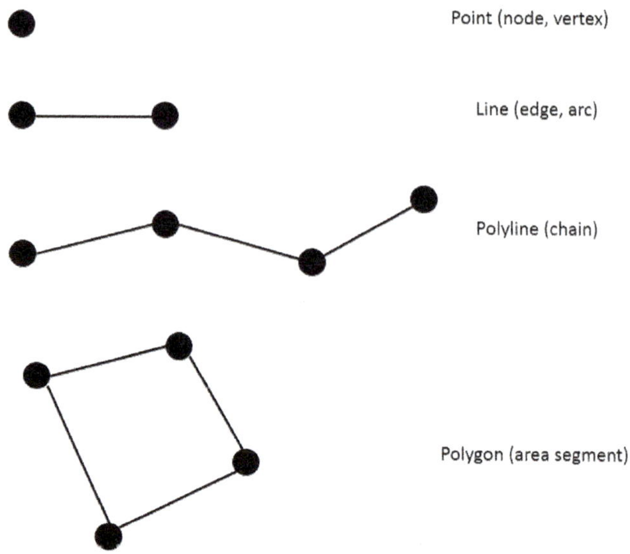

Fig. 1.6 Types of objects represented using the vector data type

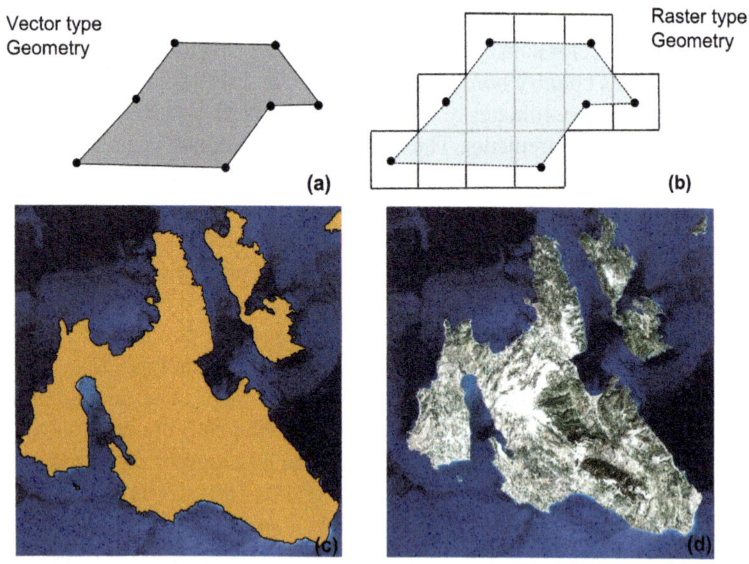

Fig. 1.7 Examples of comparison between vector (**a**, **c**) and raster (**b**, **d**) data types

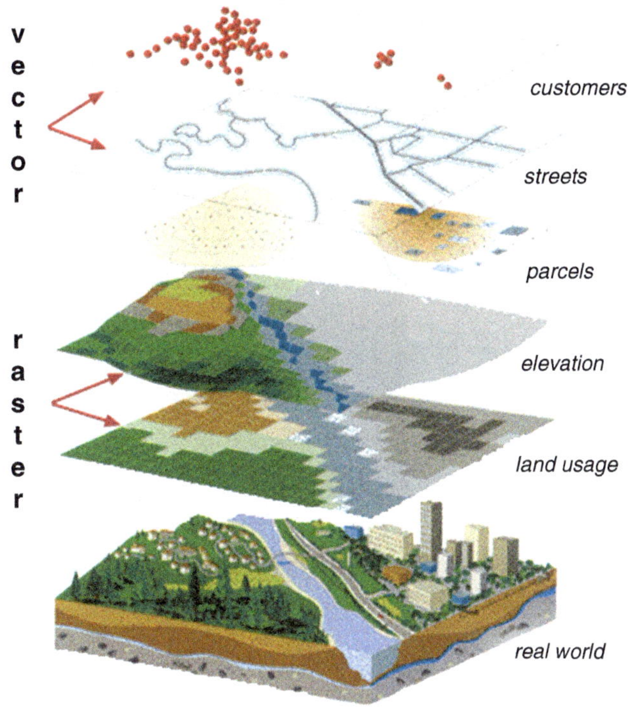

Fig. 1.8 The combination of vector and raster data files can produce representative and analytic maps of the real world (Source: http://www.seos-project.eu or http://serc.carleton.edu/eyesinthesky2/week5/intro_gis.html)

1.7 GIS Capabilities and Procedures

Main characteristics of the GIS, are the geographical dimensions and labels that are assigned to all kinds of data handled through such systems. The geographic coordinates (latitude, longitude) and the elevation define the physical objects in space (geolocation), attributing realistic dimensions to the data representation (Fig. 1.9). In this way, any unrelated kinds of data and information are related in time and space through the three previously mentioned key variables. Such characteristic capabilities have initiated a promising future for these systems and have created new horizons for application handling and visualizing of spatial information. These perspectives are enforced through the evolution of space-based navigation systems technology, such as the Global Positioning System (GPS) developed by the USA, and Galileo (a global navigation satellite system) supported by the European Union (EU) and the European Space Agency (ESA).

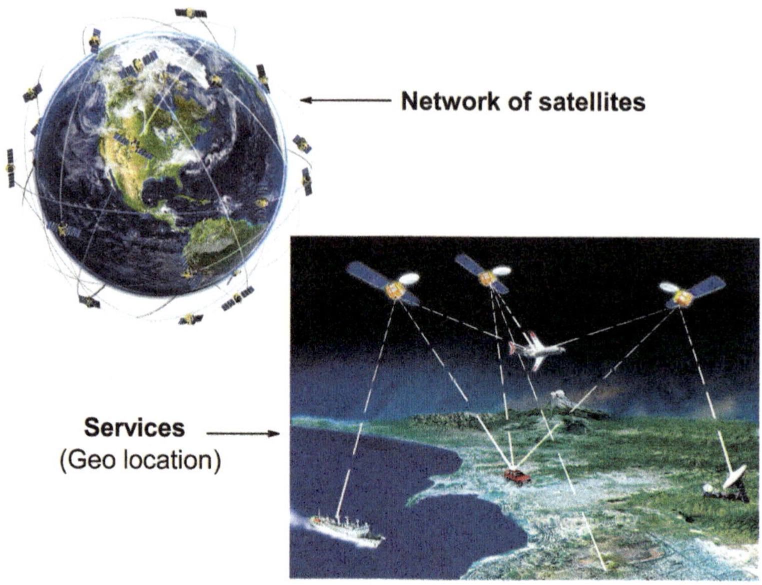

Fig. 1.9 Use of satellite networks for the provision of accurate geolocation services worldwide

Apart from the GIS capability to give geographical dimensions for any spatial information, there is a wide range of operations and procedures characterizing such systems, categorized in the following sections.

1.7.1 Input/Capture Data

In the GIS, new data are created or inputted by scanning or digitizing (Fig. 1.10) and are georeferenced. There are many tools available that facilitate the editing of data, the correction of errors, and the addition/editing/deletion of features. Digitization procedures allow capturing of objects and creation of vector files. Additionally, almost any other data types can be imported and linked to existing databases and spatial information. Then, labeling of spatial features to be identified by names, codes, or other key parameters is enabled.

1.7.2 Georeference/Georegistration

Scanned maps and images can be georeferenced through a GIS. *Georeference* is the procedure with which geographic dimensions are obtained for a digital image (or data file) containing spatial information. One image can be georegistered to various spatial reference systems (geographic coordinate systems). The procedure

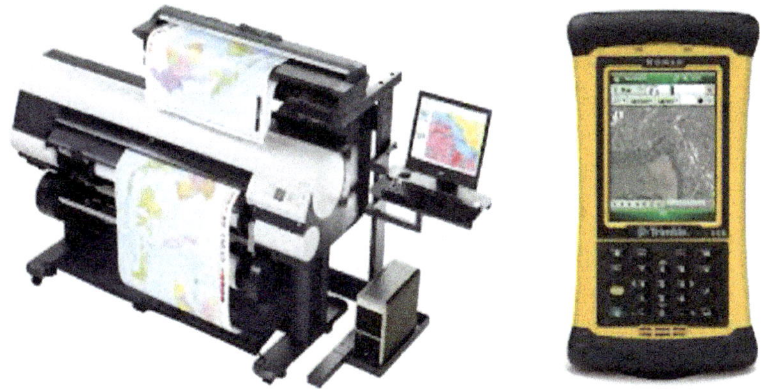

Fig. 1.10 External devices (printers, scanners, geolocating devices) that cooperate with GIS

for changing spatial reference systems requires some common steps and is very important because every pixel and area must correspond to the exact real location/area on the Earth's surface. Moreover, these procedures comprise an appropriate preprocessing step in order to spatially compare/join two (or more) image files with different spatial reference systems.

A characteristic example of georeference is given in the series of Figs. 1.11, 1.12, and 1.13. More analytically, Fig. 1.11b provides a characteristic example where the image, which shows the countries around the Mediterranean Basin and Europe, has a different projection system in relation to the image showing the coastline (Fig. 1.11a). In order to register (spatially match) the image of the countries (Fig. 1.11b) to the image of the coastline (Fig. 1.11a), which is considered the "base image," the georegistration procedure is performed.

The georegistration procedure is done by carefully selecting discrete ground control points (GCPs) that represent the same locations in both images (Fig. 1.12a, b). After the collection of GCPs, an algorithm able to resample the image is chosen so that it can be registered to the base image. The resampling method provides a final image output that can be fitted to the base image.

The final result of the georegistration is given in Fig. 1.13, which provides the image of the countries around the Mediterranean as being georegistered to the image of the coastline. Possible distortion of the image, especially at its edges, is the result of its polynomial resampling during the georegistration procedure. In Fig. 1.13 the georegistered image was overlaid onto the base image. In this example, the low resolution of the images (especially the image of the Mediterranean coastline) and the choice of the GCPs led to a matching which is in general satisfactory but lacks accuracy in some areas (for example, in the coastline of the northern part of the Black Sea).

The accuracy of the procedure is evaluated through statistic parameters, such as the root mean square (RMS) error, which refers to the distance between the same GCPs in the two examined images.

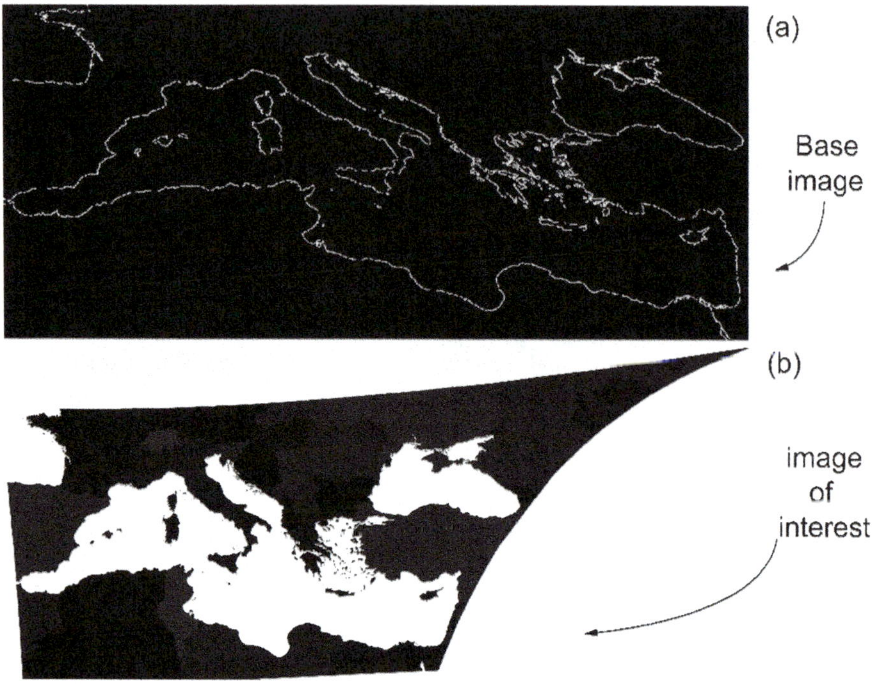

Fig. 1.11 (**a**) Image of the coastline of the Mediterranean Sea (base image). (**b**) Image of the countries around the Mediterranean that must be georegistered to the base image of the coastline

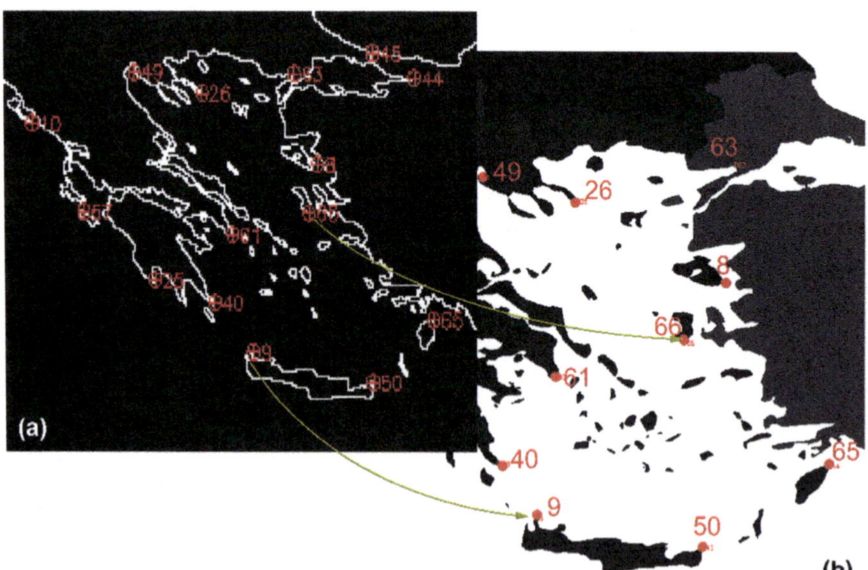

Fig. 1.12 (**a**) Image of the coastline of the Mediterranean Sea (base image) zoomed in on the area around Greece where *red points* represent the GCPs. (**b**) *Red points* represent the GCPs for the Greek periphery in the image of the countries around the Mediterranean (Fig. 1.11b)

Fig. 1.13 Zoomed-in part of the final image produced by the georegistration procedure. It can be seen that the image of the Mediterranean countries has been spatially matched with the image of the Mediterranean coastline (Fig. 1.11a)

1.7.3 Data Management

The management of data through a GIS involves procedures such as linkage among different spatial objects and linkage of attribute data with spatial objects and external databases. Modern and easy-to-handle queries help users to choose specific objects from the initial database, and ask questions using multiple criteria by comparing and integrating different layers of data. Such a visualized case is presented showing the Greek region of Central Macedonia using a cyan color in the digital map (Fig. 1.14).

1.7.4 Spatial Analysis

Buffer Zones

GIS is able to manipulate geographic data and distances, and can easily measure areal extent. Using such capabilities, buffer zones can be extracted by delimitating areas of interest around points, lines, or polygon features. Alternatively, a buffer zone represents an explicitly delimited area that surrounds an initial line/point/polygon (nodes) at a specified maximum distance from an object that is represented by one of these three nodes (e.g., Fig. 1.15).

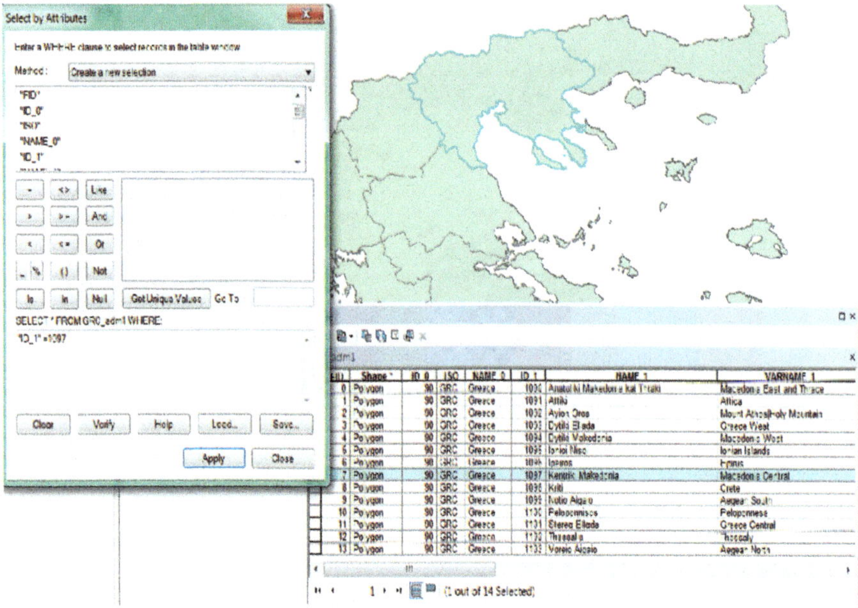

Fig. 1.14 Example of a query presenting specific spatial attributes of a data file

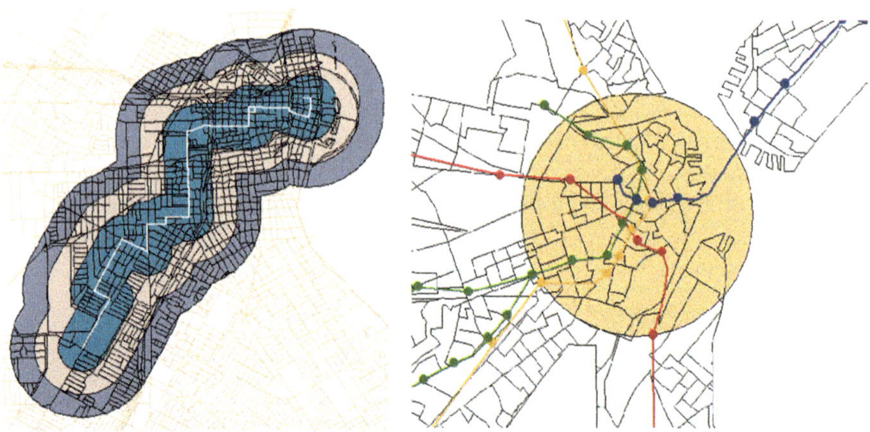

Fig. 1.15 Buffer zones at different distances from a road network (represented by a *white line*) as well as from specific points (e.g., railway stations)

Overlay/Join Different Layers of Data

As already mentioned, GIS provide a variety of tools and methods to overlay/combine/join different layers of data. In particular, the *spatial join* procedure is very useful in combining (relating) different vector data, using common records of their attributes tables (Fig. 1.16).

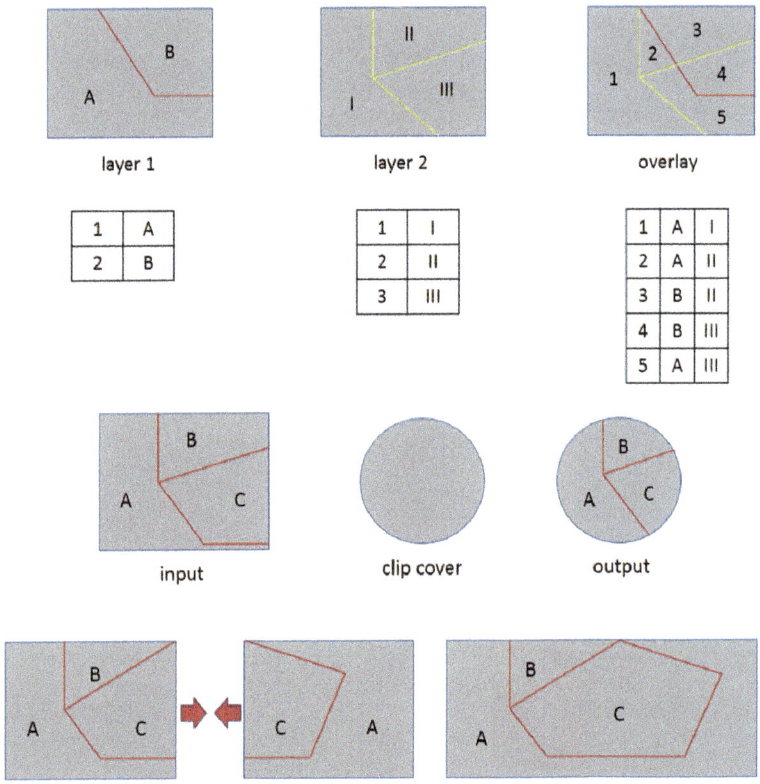

Fig. 1.16 Schematic representation of the procedure of overlaying/clipping of vector data and the "appending" of two different vector files of data

Create/Append Subsets of Data

Through GIS, a variety of tools and methods to create and append subsets of data is available. Characteristic schematic diagrams are given in Fig. 1.16.

Edge Matching/Merging/Clipping

Edge matching and merging procedures for spatial data are supported in GIS. It has to be mentioned at this point that edge matching is useful for georeference and geographic image registration purposes (Fig. 1.17).

Map Algebra

Map Algebra is an analysis language, which was introduced to perform cartographic spatial analysis of raster data (Fig. 1.18). Map Algebra uses math-like

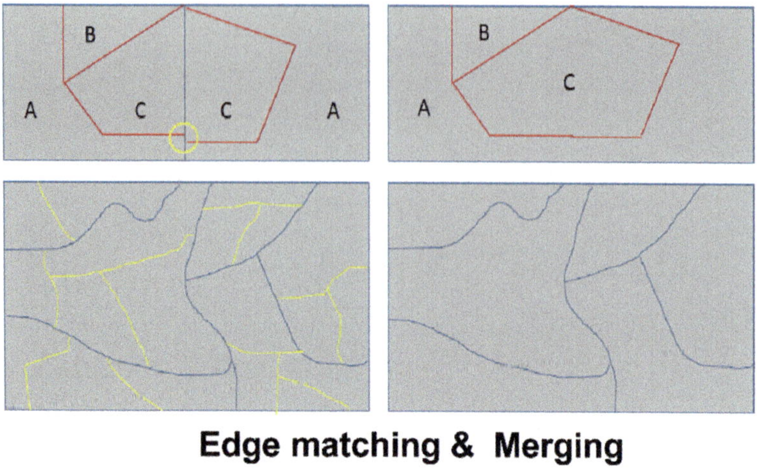

Fig. 1.17 Schematic representation of the procedure of merging/edge matching regarding two different vector files of data

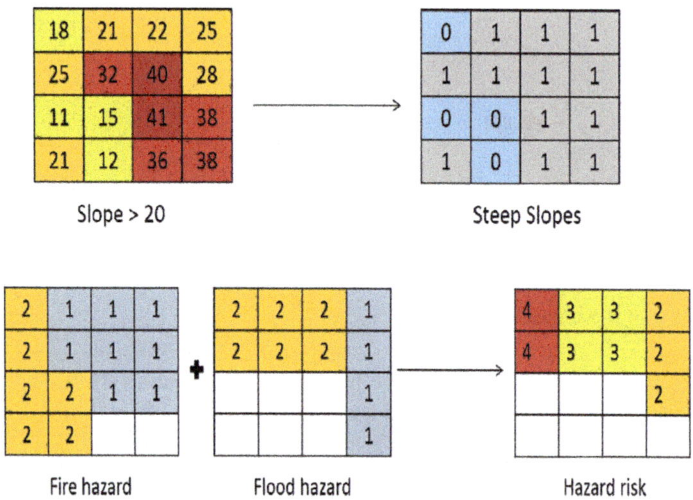

Fig. 1.18 Schemes of different data layers that can be combined with the use of Map Algebra operations to provide final two-dimensional results in raster format. In the first case, the *pixel with "1"* represents all the pixels with slope values above 20 degrees, while in the second case, the *pixel with "1"* displays the result of the total final environmental risk in an area, adding the risks of fire hazard and flood hazard

expressions containing operators and functions. More analytically, Map Algebra operators, such as arithmetic, relational, Boolean, logical, and combinatorial operators, work with one or more input data files to develop new ones.

- *Arithmetic* operators allow addition, subtraction, multiplication, and division. Arithmetic operators can also be used for unit conversion.

- *Relational* operators allow the building of logical tests, returning values of true (1) and false (0).
- *Boolean* operators, such as *AND*, *OR*, and *NOT*, allow combination of logical tests. Boolean operators return value of true and false.
- *Logical* operators *DIFF*, *IN*, and *OVER* also allow the building of logical tests on a cell-by-cell basis, but they are implemented with specific rules.
- *Combinatorial* operators combine the attributes of multiple raster data files. Such operators perform all unique combinations of values and assign a unique ID to each, which is then returned to the output grid.

A simple and characteristic example of using Map Algebra is given in Fig. 1.19. Figure 1.19a presents the regions of Cyprus as orange-colored polygons. A raster file of 30 m spatial resolution, which contains the altitude values above mean sea level, is overlaid (the dark blue regions are the more mountainous ones). Let us assume that we want to isolate regions with altitude higher than 700 m. Using a Map Algebra operation, all the pixels with values above 700 m are selected, and in Fig. 1.19b, all these pixels are colored with blue colors. All the other pixels with values below the abovementioned threshold are extracted from Fig. 1.19b.

Spatial Interpolation

The term *spatial interpolation* describes a methodology to estimate the unknown values of points by using points with known values (Fig. 1.20). More precisely, spatial interpolation estimates the unknown values of any geographic point dataset, such as elevation, precipitation, atmospheric concentrations of chemical compounds, noise levels, etc.

The interpolation constitutes a methodological category of spatial analysis, which comprises the process of managing/analyzing spatial information to extract new information from the original data.

There are many interpolation methods, divided into two main general categories: the *deterministic* and the *geostatistical.*

Deterministic In the deterministic approach, the mathematical functions that are used to calculate the values at unknown locations are based either on the degree of similarity or on the degree of smoothing in relation to neighboring data points. Characteristic examples of the determinitic methods, are the Inverse Distance Weighting (IDW) method and the Radial Basis Functions (RBF), respectively.

Geostatistical The geostatistical methods use both mathematical and statistical methods to predict values at all locations within a region of interest. These methods not only create spatial estimation and error surfaces but also produce probability output maps depending on the user's needs.

Among the geostatistical methods, one of the most well-known and widely used methods is the *Kriging method* (Krige 1951; Matheron 1963), which is a Gaussian

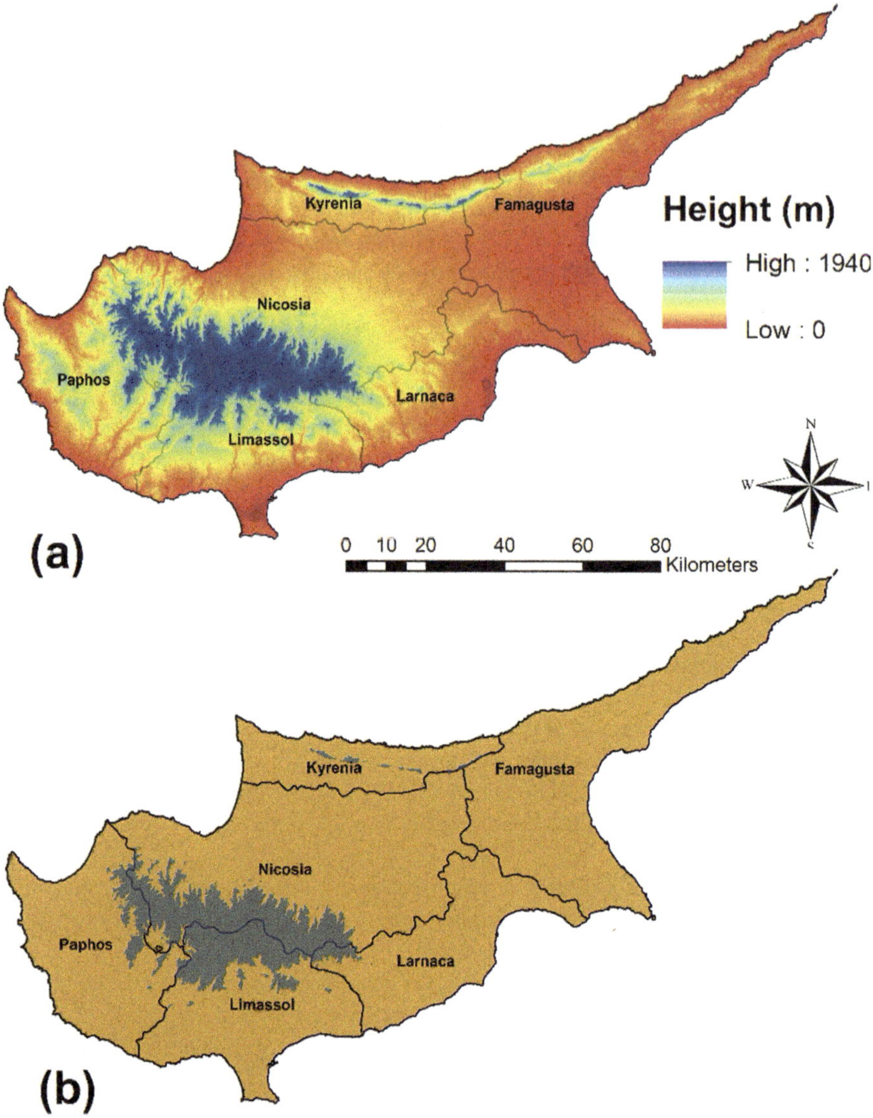

Fig. 1.19 Example of making thematic digital maps using Map Algebra. (**a**) The island of Cyprus. The *polygons* represent its main regions, while the *colored surface* represents the altitude in 30 m spatial resolution. (**b**) The *gray-colored regions* highlight all the pixels with altitude values above 700 m

regression process of interpolation. As in many geostatistical models, the sampled data are interpreted as the result of a random process. Depending on the stochastic properties of the random field and the various degrees of stationarity that are assumed, different methods for calculating the weights can be deducted and different types of Kriging can be applied (for example, simple Kriging, ordinary Kriging, universal Kriging, etc.). To make an estimation of an unknown value at a specific

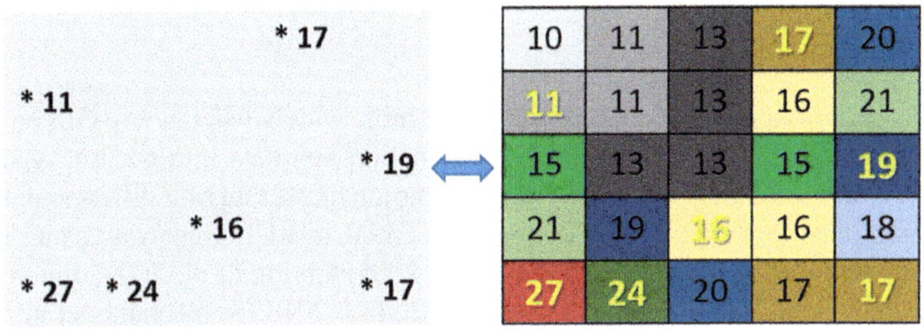

Fig. 1.20 *On the left*, a point dataset of known values. *On the right*, a raster file generated by the implementation of spatial interpolation at the initial points. Unknown values are estimated with a mathematical formula based on the values of nearby known points

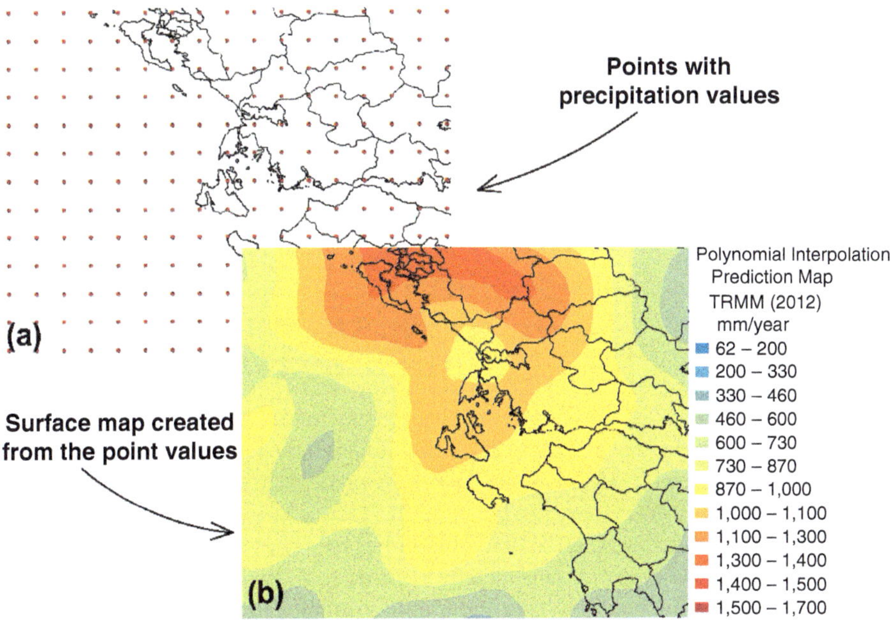

Fig. 1.21 (**a**) Point values used to create a continuous surface (estimation map) of values, from the initial point values. (**b**) Creation of a continuous surface (estimation map) from point values

location, Kriging uses variograms for the quantification of the spatial structure in the data and the actual values of the measured sample points around the location.

As an example, Fig. 1.21a gives a sample of point data, where the annual precipitation for the year 2012 in the western part of the Greek periphery is provided. These point data come from the Tropical Rain Measurement Mission (TRMM) satellite, and they represent the centers of grid cells of spatial resolution of 0.25 degrees. In order to create a continuous surface that provides spatial estimates among all the points of the area of interest, an interpolation methodology is performed. The result of this interpolation can be seen in Fig. 1.21b.

1.7.5 Classification

The procedure of classification is one of the most widely used in any GIS environment as it helps to categorize spatial information and data into general types and compare data and images per pixel using common classes among different datasets. The classification procedure is not only essential in LULC analysis using multispectral satellite imagery (e.g., Yuan et al. 2005; Liying et al. 2009; Julien et al. 2011; Petropoulos et al. 2011; Kolios and Stylios 2013; Symeonakis et al. 2011; Tzanopoulos and Vogiatzakis, 2010; Gounaridis et al. 2014) but also very important in land degradation, desertification, and many other physical phenomena and procedures with spatial dimensions (e.g., Koulouri and Giourga 2007; Hill et al. 2008; Symeonakis et al. 2014; Koukoulas and Blackburn 2005).

Classification is the process of data analysis that arranges data according to the values of information, allowing the recognition of discrete spatial patterns in the data. The basic types of data classification are:

- Equal interval classification
- Quantile classification
- Jenks natural breaks classification
- Geometric interval classification
- Standard deviation classification

It is mentioned that there are two general types of classification techniques for digital images: the *supervised* and the *unsupervised* classifications.

In the *unsupervised classification* the pixels are categorized based on their multispectral properties. These categories are known as *clusters*. The user identifies the number of clusters and the bands of multispectral imagery. Using such information, the image classification algorithm (classifier) identifies clusters. Some well-known image clustering algorithms are *K-means* and *ISODATA*.

In the case of *supervised classification,* the user selects the main representative samples for each LULC class in the digital imagery. The selected samples of each recognized LULC cover class is called the *training sample* of the class. Then the image classification software uses the training samples to identify the land cover classes in the entire image.

The classification of LULC is based on the spectral signature defined by the training sample of the class. The classification algorithm determines each class based on the training samples. The most common supervised classification algorithms are the *maximum likelihood* and *minimum distance* classifications, but nowadays there is a variety of modern classifiers, such as the *support vector machine* and various types of *artificial neural network.* A representative example of a supervised classification result is given in Fig. 1.22.

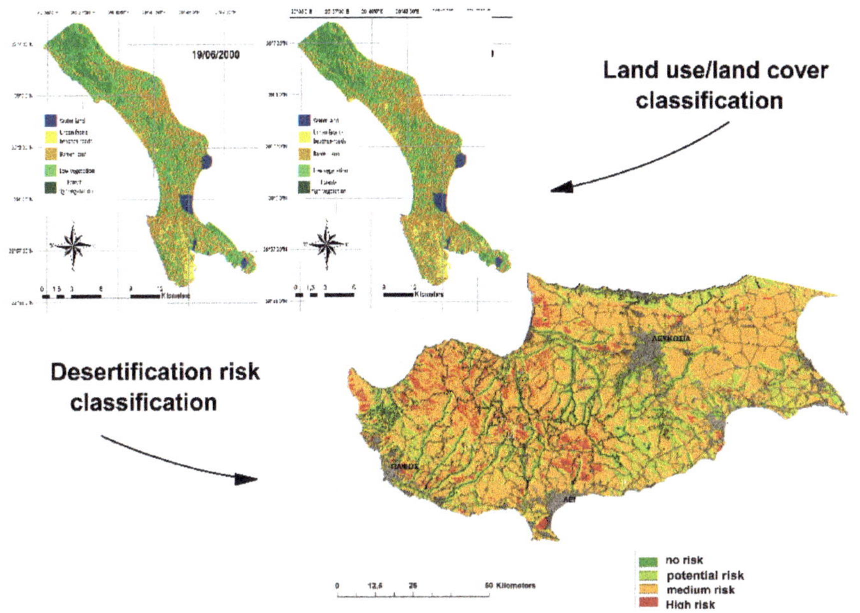

Fig. 1.22 Example of a supervised classification result in a satellite imagery (Landsat) for land use/land cover analysis

1.7.6 Modeling

The modeling in GIS is usually used to identify or predict a process that creates a certain spatial pattern. Some indicative modeling applications include diffusion (e.g., in which direction and to which extent is atmospheric pollution spreading?), interaction issues (e.g., where do people migrate to?) and "what-if" scenarios (e.g., if the dam is built, how many people will be displaced?) (e.g., Goodchild and Proctor 1997; Ball and Luk 1998; Maguire et al. 2005; Brown et al. 2006; Bhatt et al. 2014; Louca et al. 2015; Armstrong et al. 2015). The various GIS software programs are based on compatible programming languages as well as supporting extensions (*plug-ins*) that can be used to simulate the spatiotemporal evolution of parameters and phenomena.

1.7.7 Display/Export

The majority of the GIS offer innovative advanced visualization and export capabilities. More specifically, data from multiple layers can be combined and visualized, graphs and statistics can be overlaid, and digital maps (even 3D) with legends and scale bars can be created and exported in several image data types (e.g.,

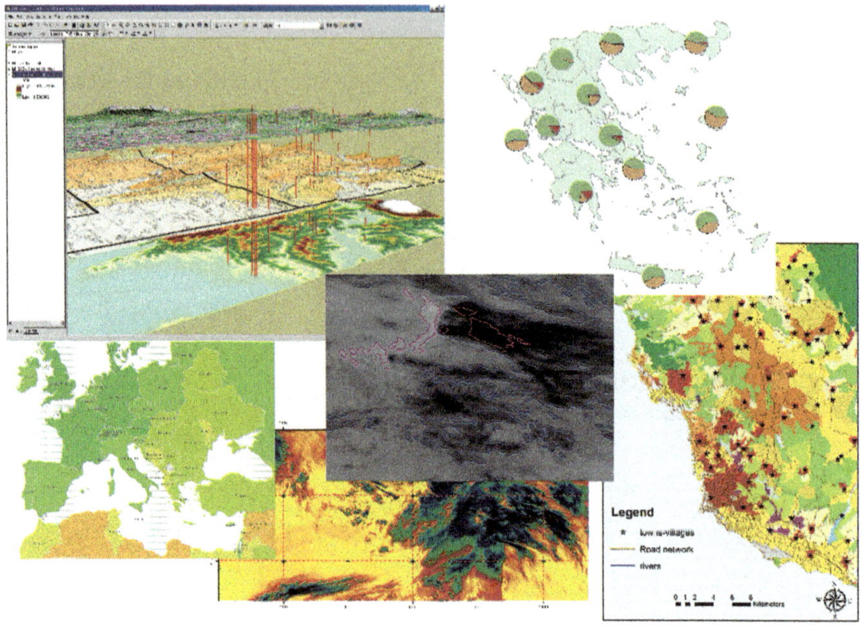

Fig. 1.23 Examples of GIS graphical results

Papakonstantinou et al. 2011, 2013; De Kluijver and Stoter 2003; Berry et al. 2011; Brown et al. 2006). Figure 1.23 illustrates some characteristic GIS visualization examples.

1.7.8 Programming Languages in GIS

All modern GIS allow users to make their own procedures and analyses using scripting programming, which requires a compiler for the respective programming language. Most GIS software nowadays supports scripting through the C, C++, Visual Basic and Python programming languages.

1.8 WebGIS Definition and Overview

The last decade has witnessed rapid growth of development and operation of Web geographic information systems (WebGIS). GIS have been extensively applied to many scientific, commercial, and governmental sectors as spatial data are inextricably linked to a specific location in the form of street addresses, postal codes, or latitudes and longitudes. With the spread of the World Wide Web, a new branch of

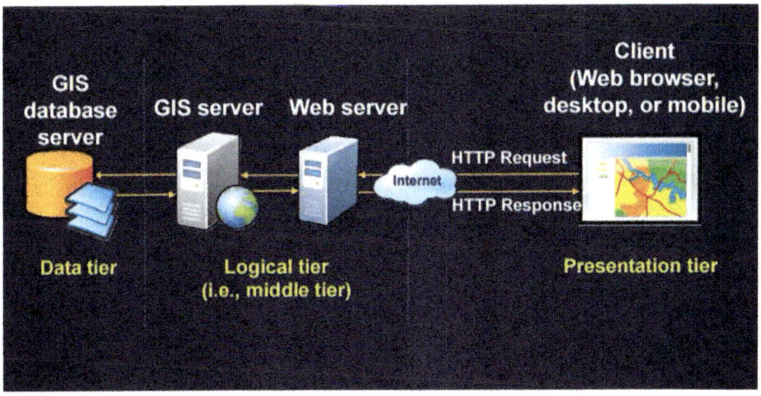

Fig. 1.24 Basic components of a WebGIS application

GIS—WebGIS—has gradually started to grow. The GIS data and their operations have been made available online, serving as modern cartographic tools that facilitate the creation of digital thematic maps and examination of the impacts of changes to the maps interactively (e.g., Rigaux et al. 2002; Longley et al. 2010).

In comparison with desktop GIS, WebGIS supplies a more efficient platform for timely (sometimes real-time) integration and dissemination of georeferenced data, thereby enabling efficient data use and effective public health interventions. Furthermore, the broad available bandwidth and the increasing quality of wireless connections offered by the mobile telephone companies help with the promotion and usability of the WebGIS applications.

Trying to provide a short definition of what a WebGIS is, it can be said that *a WebGIS is any GIS that uses Web technologies and operates on the World Wide Web*. There are three main parts (tiers) that comprise the basic architecture of a WebGIS (Fig. 1.24): the presentation tier, the logic tier and the database tier (e.g., Frehner and Brandli 2006; Fu and Sun 2010).

- The *presentation tier* serves as a graphical user interface (GUI) to present the result of spatial data, allowing end users to interact with the back-end services. The presentation tier can be a desktop or a mobile device, but in the WebGIS applications the presentation of any kind of data and information is shown through a Web browser.
- The *logic tier* communicates with multiple data sources and interacts with end users to analyze and manage/analyze data coming from the services from the data provider. In the case of a WebGIS, the application server is, practically, a Web server in cooperation (or integrated) with a GIS server.
- The *data tier* is a provider for remotely sharing data and information. Each data provider service offers a set of interfaces through which client applications can exchange and manage remote data.

The main benefit of this three-component structure is the separation of the application into different tiers where every tier operates independently, thus allowing easier management of any system update, change, or even failure.

The *presentation tier* constitutes the interface from the user side and is accessed directly by users. On the other hand, it communicates with the logic tier (Web server), forwarding the client request and displaying the data/information that respond to the client's request in the form of Web page appearance. All the Web pages of such applications have been developed based on HTML and JavaScript.

The *logic tier* has its own layer; it controls all the application's functionalities and it processes the client requests. This layer is usually based on an Apache HTTP server. The interaction language is PHP, which is a scripting language widely used for Web development and also used as a general-purpose programming language.

Finally, the *data tier* includes the data persistence mechanisms; it consists of the database server where files are uploaded and there is a data access layer that encapsulates the persistence mechanisms and exposes the data. The data access layer provides an application programming interface (API) interacting with the logic tier in order to provide methods for managing the stored data without causing dependencies on the data storage mechanism. In this layer, a database server is used and configured properly to handle the data.

An overall view of the WebGIS application components, user requests and interactions among the system components is given in Fig. 1.25, which presents user interactions via the Web browser with the WebGIS.

More analytically, the Web browser makes requests (driven by the user) to the server using the HTTP protocol. The server forwards the requests and accesses data from the database, and then it returns the requested information by generating Web pages serving as responses to the user's requests. The Web browser renders the responses from the server and the results appear on the user's screen.

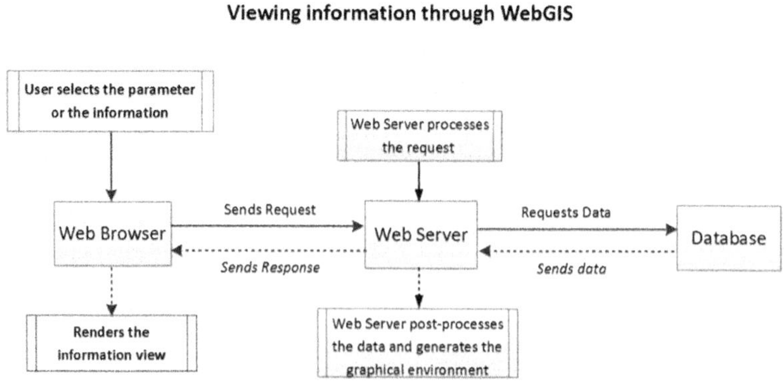

Fig. 1.25 Information flow in the life cycle of a client request (Kolios et al. 2015)

1.9 Web Mapping: Definition and Types of Web Maps

An essential part of the WebGIS applications is the Web mapping and the various types of Web maps. Web mapping is the process of viewing and extracting information from maps provided by a GIS. A Web map is an interactive digital object. A WebGIS application provides such Web maps and end users can access all the capabilities and the features of the Web mapping service.

There are several types of Web maps. A general categorization includes the *static* and the *dynamic* Web maps (Kraak and Brown 2001). Nevertheless, the advances in Web mapping have already led to many different subcategories. Those most commonly used are provided in Table 1.1.

Except for the static type, all other types of Web maps included in Table 1.1 can be considered as interactive maps. The term *interactive* implies that the end user can somehow interact with the map (for example, by selecting different map data layers to view, or by zooming in on a particular part of the map). When interacting with the relevant Web page, the map image is automatically updated. The general flow of procedures starts from the user who asks to view some information from a Web map of the Web server. The server passes the request to the Web mapping server, which then transfers together all the data according to the request. The map is passed all the way back to the end user's Web browser (Fig. 1.26).

1.10 Geospatial Web Map Services

The Web mapping of spatial data and information, which is provided by a GIS, needs the relevant geospatial Web Map Services (WMS). WMS is a standard protocol for serving georeferenced map images over the World Wide Web that are generated by a map server using data from a GIS database. It is good to mention at this point that WMS is not a Web site and it includes services that can be called to return information. The request to a Web service for attribute information for a specific point is referred to as Web Feature Service (WFS) and allows a client to retrieve and update geospatial data encoded in Geography Markup Language (GML) format.

The WMS standard specifies a number of different request types, but two of them are essential to any WMS server:

- The *GetCapabilities* request, which returns parameters about WMS (such as the map image format and WMS version compatibility) and the available layers (map bounding box, coordinate reference systems, Uniform Resource Identifier (URI) of the data and whether the layer is mostly opaque or not). This information practically returns an XML document with metadata on the Web map server's information.

Table 1.1 Major types of Web maps

Type of Web map	Short description
Animated Web maps	Animated Web maps can provide changes over time by animating one of the graphical or temporal variables. Various data, multimedia formats, and technologies allow the display of animated Web maps, such as SVG, Adobe Flash, Java, etc., with varying degrees of interaction. Examples of animated Web maps are weather maps and maps displaying dynamic natural or other phenomena (such as water currents, wind patterns, etc.). This type of Web map can be used even in real-time (or near-real-time) applications; the provided data and information are collected by sensors and the maps are generated or updated at regular intervals or on demand.
Collaborative maps	Collaborative maps have an increasing potential and this type of map is created and improved with the contribution of various people and sources through the Web. Technically, relevant applications allow simultaneous editing across the Web, but it has to be ensured that geometric features are edited by one person, fulfilling specific standards of accuracy, and cannot be edited by other persons at the same time. Some collaborative Web mapping projects are:
	Google Map Maker
	OpenStreetMap
	WikiMapia
Customizable Web maps	Customizable Web maps are practically the final products of complex Web mapping processes offered though APIs, and they can be reused in other people's Web pages and products. Characteristic examples of customizable Web maps are the Open Layers Framework, Yahoo! Maps, and Google Maps.
Distributed Web maps	Distributed Web maps are created from a distributed data source. The Web Map Services (WMS) protocol offers a standardized method to access maps on other servers. WMS servers can collect the different sources of information, reproject the map layers (if necessary), and send them back as a combined image containing all requested data layers. For example, one server may offer a topographic base map and other servers may offer thematic layers.
Dynamically created Web maps	Dynamically created Web maps are created on demand each time the user reloads the Web pages, often from dynamic data sources, such as databases. The Web server generates the map using a Web map server or self-written software.
Static Web maps	Static Web pages are view-only with no animation and interactivity. They are created once, often manually, and are infrequently updated. Examples of data and graphics formats used for static Web maps are PNG, JPEG, GIF, or TIFF for raster files, and SVG, PDF, or SWF for vector files. Often, these maps are digital files of scanned paper maps.

- The *GetMap* request, which returns a map image according to the user's request. The parameters that accompany the map image include the width and height of the map, coordinate reference system, rendering style, and image format.

More precisely, a WMS standard provides a simple HTTP interface for requesting georeferenced map images from one or more distributed geospatial

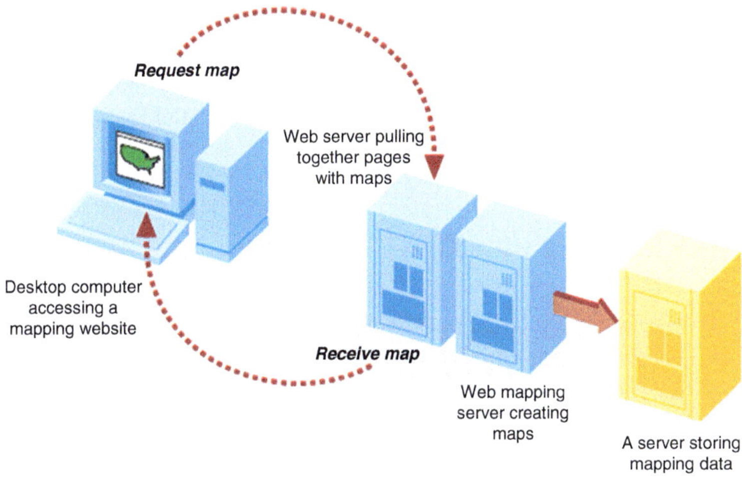

Fig. 1.26 Schematic flow diagram of interaction among a mapping Web site, the end user, and the back-end programs running on the servers (Source: http://www.safaribooksonline.com/library/view/web-mapping-illustrated/0596008651/ch01s03.html)

databases. A WMS request defines the geographic layer(s) and the area of interest to be processed. The response to the request is one or more georeferenced map images (in a wide variety of data formats, such as BMP, JPG, TIF, etc.) that can be displayed in a Web browser application. The interface also supports the ability to specify whether the returned images should be transparent so that layers from multiple servers can be combined or not.

Furthermore, a WMS-compliant server can handle types of WMS requests, like:

- *GetFeatureInfo*, which return information about attribute(s) at a query (mouse click) location in a map.
- *DescribeLayer*, which returns the feature types of the specified layer or layers that can be further described using WFS requests. This information practically returns an XML description of one or more map layers.
- *GetLegendGraphic*, which returns a legend image (icon) for the requested layer, with label(s).

At this point, it is worth pointing out an alternative service to WMS, the Web Coverage Service Interface Standard (WCS). WCS defines Web-based retrieval of coverages for digital geospatial information that represents space/time–varying phenomena useful for client-side rendering, as input into scientific models, and for other clients. WCS uses the coverage model of the OGC GML Application Schema for Coverages (Baumann 2012).

Like the WMS and WFS service instances, WCS allows clients to choose portions of a server's information holdings based on spatial constraints and other query criteria. Nevertheless, one basic difference between WCS and WFS is that WCS provides available data together with their detailed descriptions, while WFS portrays spatial data to return static maps (rendered as pictures by the server). Additionally, WCS returns coverages representing spatiotemporally varying parameters that relate a spatiotemporal domain to a (possibly multidimensional) range of properties, while WFS returns only discrete geospatial features.

1.11 Web Mapping with Tiles

A tiled Web map is an interactive map displayed in a browser by joining dozens of image files individually requested from a Web server (e.g., Fig. 1.27). Nowadays, Web mapping using tiles is considered the most popular way to display and navigate maps, gradually replacing any other method, such as WMS, which typically displays a single large image, with arrow buttons to navigate to nearby areas. There are several major advantages to tiled maps, among which are the following:

- Each time the user pans, most of the tiles are still relevant, and they are kept displayed, while new tiles are fetched. This greatly improves the user experience, compared with fetching a single map image for the whole view.
- Displaying rendered images requested from a Web server is much less computationally demanding than rendering images in the browser (a benefit over technologies such as WFS).
- Tiled maps are simple to use. The coordinate scheme describing map tiles is simple, making it easy to implement integrating technologies on the server, Web, desktop, and mobile devices.

In conclusion, the tiles divide a map into different subregions (the division is dependent on the zoom level) and the browser each time shows a small piece of a map (according to the zoom level). This technique makes the map tiles a computationally modern and quick way to show interactive maps, allowing integration

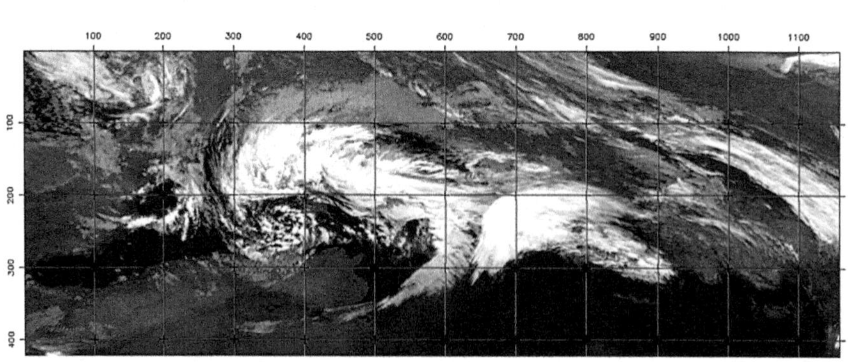

Fig. 1.27 Example of a satellite image in map tiles

with other Web technologies in order to show information and plot spatial/nonspatial information on the server, Web, desktop, and mobile devices.

1.12 Application Programming Interface (API)

The application programming interface (API) is a set of routines, protocols, and tools for building software applications. The API defines how software components can interact while APIs are used when programming graphical user interface (GUI) components. APIs help the design and development of a program by providing all the building blocks required by a programmer. APIs often come in the form of a library, which includes specifications for routines, data structures, object classes, and variables.

During the development of any WebGIS application, modern APIs are used to provide and improve many functionalities and operations regarding analysis, management, and visualization of spatial data. Characteristic categories of such APIs are briefly described in Sects. 1.12.1, 1.12.2, and 1.12.3.

1.12.1 Ajax/JavaScript APIs

Ajax (asynchronous JavaScript and XML) is a term that describes an approach of using a combination of a number of existing technologies, like HTML, JavaScript, etc. The jQuery (http://jquery.com) library is a full suite of Ajax capabilities. These functions/methods/applications allow the user to load data from the server without a browser page refresh and are primarily used for Web browser development. As a result, they are well suited to Web mapping applications with GIS. There is direct support for KML and GeoRSS file types as well as raster tile layers. Many user interface (UI) features (like panning/zooming, etc.) are part of these APIs. It is noted that JavaScript is compatible with any Web server technologies, making the development and the implementation very flexible.

Nevertheless, JavaScript lacks advanced UI features, such as animation, rich graphic design, transparency, etc. For Web mapping applications in GIS, the JavaScript APIs provide vector and raster overlay import, suitable for widespread publication on every type of browser.

1.12.2 SOAP/XML APIs

The Simple Object Access Protocol (SOAP) is a protocol specification for exchanging structured information in the implementation of Web services in computer networks. The SOAP/XML Web service APIs support server-side calls. Requests can be made with a great variety of result objects in XML. These APIs offer ease of use With the .NET framework and other development technologies that support SOAP/XML. They run as multithreaded background services (e.g., bulk geocoding) and support any type of UI, including older browser versions, mobile devices, and smart clients. Their main disadvantages are that they return static maps and there are no navigation controls, so these features need to be developed as part of the application.

1.12.3 Silverlight APIs

These APIs use Microsoft Silverlight technology, which are primarily .NET-based APIs with code that runs inside the user's browser. Their advantages include ease of development with .NET and powerful UI features, such as animation, rich graphic design, and transparency. In addition, they manage client-side code means with high-performance display of large vector datasets and, in some cases, GPU accelerated graphics. Applications can use combinations of Map APIs, i.e., Silverlight for visualizing data and SOAP/XML for geocoding.

1.13 Modern WebGIS Architectures

A modern WebGIS application constitutes a software framework that integrates modules of software providing generic functionality along with specific user-written code to achieve the system functionalities. A WebGIS application can utilize various open-source frameworks and libraries to achieve a professional level of development and maintenance. For example, the *Codeigniter PHP MVC Framework,* the *Bootstrap CSS Framework,* the *jQuery JavaScript Library,* the *jQuery UI JavaScript GUI Library*, and the *Google Maps API Web mapping service* can be used to provide a modern Web-based GIS integrated application. Many of these tools are used in various applications (Pascual et al. 2012).

Figure 1.28 provides a schematic diagram of the components of the WebGIS architecture and their main interactions. More analytically, the user requests specific information from the server through the Web browser. The server interacts with the database to obtain the relevant data/information and then generates Web pages that are pushed to the Web pages in order for the user to see them.

In the following sections, short descriptions for the abovementioned main components of the WebGIS architecture are given.

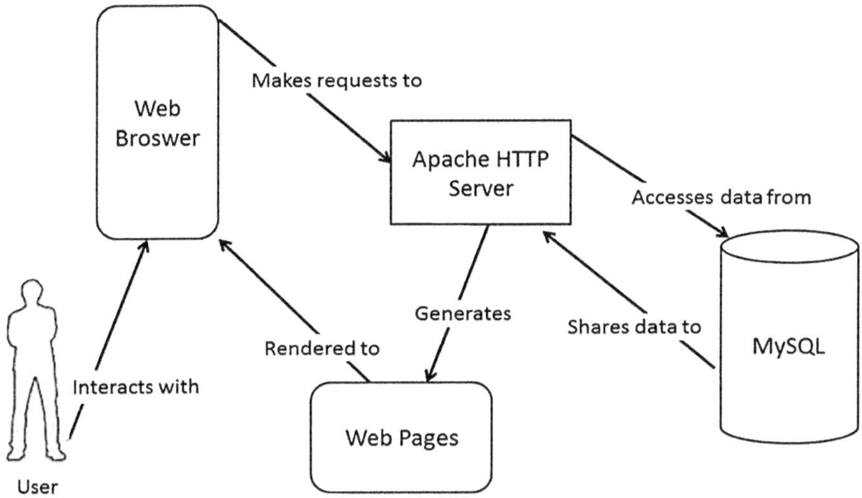

Fig. 1.28 Schematic diagram of WebGIS architecture components and their interactions

1.13.1 Codeigniter PHP MVC Framework

Model–View–Controller (MVC) is a software architectural pattern for implementing UIs (Buschmann et al. 2001). It divides a given software application into three interconnected components, so as to separate internal representations of information from the way that information is presented to or accepted by the user.

The central component of the MVC, the *model*, contains the core functionality and data. The *view* is the second component, which can be described as any output representation of information, such as a text, images, etc. Multiple views of the same information are possible, such as a bar or options for navigation or other operations. The third component, the *controller*, accepts input and converts it to commands for the model or view.

In addition to the division of the application into three components, the MVC design defines the interactions between them as follows:

The *controller* can send commands to the model to update the model's state (e.g., editing a publication document). It can also send commands to its associated *view* to change the view's presentation of the model (e.g., the user can see possible changes in a document).

The *model* notifies its associated *views* and *controllers* when a change in its state occurs. This notification allows the *views* to produce an updated output and the *controllers* to change the available set of commands. In some cases, an MVC implementation might, instead, be "passive," so that other components must inform the *model* for updates rather than being notified.

A *view* requests information from the *model*, which is necessary for generating an output representation to the user (Fig. 1.29).

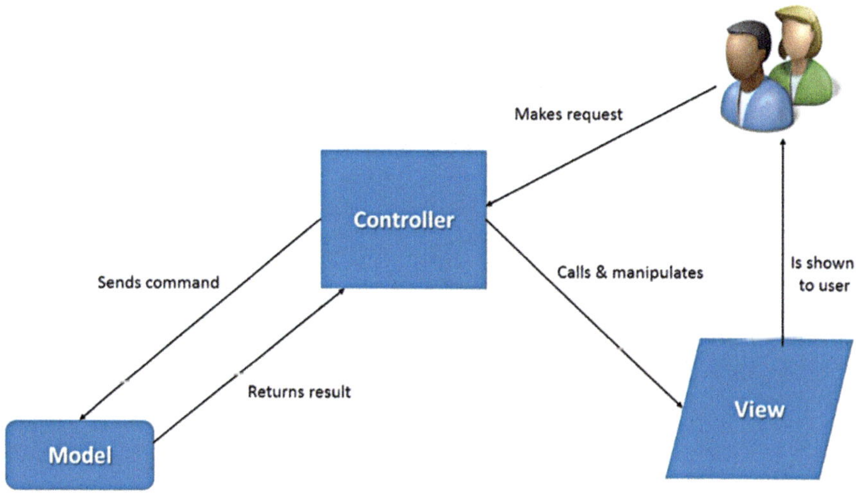

Fig. 1.29 Interactions in the Model–View–Controller Framework

1.13.2 Bootstrap CSS Framework

Cascading Style Sheet (CSS) frameworks are preprepared software modules that allow easier, more standards-compliant Web design using the CSS language. Most of these modules contain at least a grid. More functional modules also come with more features and additional JavaScript-based functions, but are mostly design oriented and unobtrusive. This differentiates CSS from functional and full JavaScript frameworks. The CSS frameworks offer different modules and tools:

- Reset stylesheet
- Grid for responsive Web design
- Web typography
- Set of icons in sprites or icon fonts
- Styling for tooltips, buttons, elements of forms
- Parts of GUIs, i.e., accordion, tabs, slideshow, or modal windows
- Equalizer to create equal height content
- Often used CSS helper classes (left, hide)

Bootstrap is an open-source CSS framework compatible with the latest versions of all major browsers. It supports responsive Web design, which means that the layout of the Web pages adjusts dynamically, taking into account the characteristics of the device used (desktop, tablet, mobile phone).

Bootstrap is used in the TEN ECOPORT and PORTWEATHER WebGIS platforms analytically presented in Chap. 2, as well as in the Geomagnetic Calculator (WMM2015) WebGIS tool presented in Chap. 4.

1.13.3 jQuery JavaScript Library

jQuery (http://jquery.com) is a cross-platform JavaScript library designed to simplify the client-side scripting of HTML. It was released in January 2006 and it is used by over 80% of the 10,000 most visited Web sites. It is the most popular open-source JavaScript library in use today.

jQuery v1.11 is used in the TEN ECOPORT and PORTWEATHER WebGIS platforms as well as in the Geomagnetic Calculator (WMM2015) WebGIS tool.

1.13.4 jQuery UI JavaScript GUI Library

jQuery UI is a collection of GUI widgets, animated visual effects, and themes implemented with jQuery, CSS, and HTML. Both jQuery and jQuery UI are open-source software under the MIT license.

1.13.5 Google Maps API

Google Maps API is a service (launched by Google in 2005) to allow developers to integrate Google Maps into their Web sites. By using Google Maps API, it is possible to embed Google Maps into an external Web site, where specific data can be overlaid. It is a JavaScript API used by over one million Web sites, making it the most widely used API for Web application development.

Google Maps API is used for the development of all the WebGIS applications analytically presented in Chaps. 2 and 4.

1.14 Modern Approaches for Developing a WebGIS Application—Cloud GIS

Nowadays, the importance of and continuously increasing needs for detailed spatial information in combination with Web technologies provide modern solutions for Web mapping using plug-ins, extensions of GIS software, and cloud computing. Especially the perspectives of cloud computing constitute one of the most significant issues in spatial information technology. It is worth pointing out that the technology and the architecture that cloud services and deployment models offer constitute key areas of research and development of GIS technology.

Cloud computing is a model for enabling ubiquitous, convenient, on-demand network access to a shared pool of configurable computing resources (e.g., networks, servers, storage, applications, and services) that can be rapidly provisioned and released with minimal management effort or service provider interaction (Grance and Mell 2011). In other words, cloud computing refers, in general, to the recent advancement of distributed computing by providing on-demand network access to a shared pool of configurable computing resources (e.g., networks, servers, storage, applications, and services) with minimal management effort or service provider interaction (e.g., Mell and Grance 2009; Muzafar Ahmad Bhat et al. 2011).

In parallel, it is already known that cloud computing can provide potential solutions to solve geospatial science challenges (e.g., Huang et al. 2010) with on-demand access to massively pooled, instantiable, and affordable computing resources. The close relationship between cloud computing and geospatial sciences, as well as the fact that the convenience and budget and energy consumption efficiencies of cloud computing have already been proven (Lee and Chen 2010), can play a significant role in the future of computing infrastructure for supporting geospatial sciences (e.g., Yang et al. 2011; Muzafar Ahmad Bhat et al. 2011).

Cloud GIS provides great computing and/or storage capacity, provisional parallel processing, and an on-demand platform for any GIS service (Fig. 1.30).

More specifically, cloud GIS uses the concept of operating GIS software and its related services on a cloud-based infrastructure and providing all the GIS capabilities through the Web, where the cloud GIS applications operate at a shared data center accessed via the Internet. Moreover, it supports virtualization technology, which allows conversion of one server into many virtual machines, thereby eliminating client–server computing with single-purpose systems. Nowadays, there are many providers who offer cloud GIS platforms (free of charge or not).

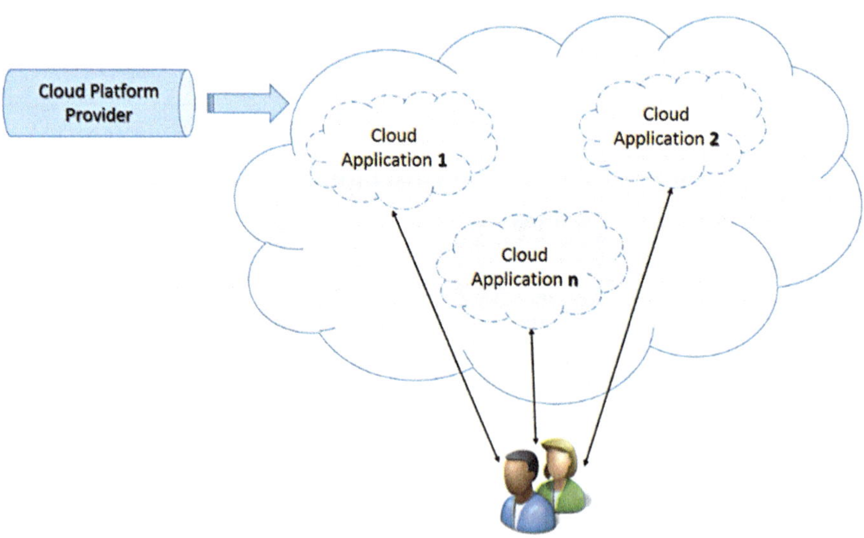

Fig. 1.30 General scheme of cloud-based services provided to end users

In addition, there are many products (extensions/plug-ins) designed to provide a GIS Cloud data layer publisher for well-known GIS applications, such as ArcMap and QGIS. Chapter 2 includes in Sect. 2.3 an example of a cloud-based GIS application covering the need to record and analyze cases of extreme weather events occurring in the Mediterranean area, which has been developed with an open-source QGIS application and includes GIS Cloud plug-ins.

1.14.1 Cloud Service Models

There are four main service models in the cloud computing environment.

Software as a Service (SaaS) This is a software licensing and delivery model where software is licensed on a subscription basis and is centrally hosted. SaaS is typically accessed by users using a thin client via a Web browser. The SaaS model provides a way of delivering applications as a service over the Internet. Instead of installing, managing, and maintaining software and hardware, one can simply access it via the Internet. The SaaS model is becoming an increasingly prevalent delivery mechanism as underlying technologies that support Web services and service-oriented architecture (SOA) mature and new developmental approaches (such as Ajax) are being increasingly used.

Platform as a Service (PaaS) This is a cloud computing model that shares applications over the Internet. It includes an application platform as a service provider on which developers can build and deploy custom applications. A PaaS provider hosts the hardware and software for its own infrastructure. The capabilities provided to the consumers with a PaaS model service are deployed onto the cloud structure where applications are created and/or acquired using programming languages, libraries, services, and tools. The consumer does not manage or control the underlying cloud structure, such as network, servers, operating systems, or storage, but he has control over the deployed applications and the configuration settings for the application-hosting environment (Grance and Mell 2011). PaaS is very useful for developers as it helps them develop and test applications without having to worry about the underlying infrastructure.

Infrastructure as a Service (IaaS) This is a standardized, highly automated cloud computing service model in which a third-party provider hosts hardware, software, servers, storage, and other infrastructure components on behalf of its users (the computing resources are complemented by storage and networking capabilities owned and hosted by the service provider and offered to customers on demand). Users are able to self-provision this infrastructure, using a Web-based GUI that serves as an IT operations management console for the overall environment. API access to the infrastructure may also be offered as an option. With the IaaS model the main capability provided to the consumer is to provision processing, storage, networks, and other fundamental computing resources where the consumer is able to deploy and run arbitrary software, which includes operating systems and applications. The consumer does not manage or control the underlying cloud infrastructure but he has control over operating systems, storage, and deployed applications; and possibly limited control of select networking components (Grance and Mell 2011). Conclusively, IaaS is considered very helpful for cloud GIS applications as it provides virtualized computing resources over the Internet.

Data as a Service (DaaS) This is based on the concept that data can be provided to the user on demand, regardless of any geographic or organizational separation of provider and consumer. This service model is closely related to SaaS and supports data access and utilization as well as data processing on demand for end users (Olson 2010). Thus, DaaS is essential to geospatial sciences, with cost effectiveness and data quality to be considered its main advantages.

Cloud computing infrastructures (platforms of services) can be either public or private. The public cloud is the most widely used and it refers to the structure and the applications owned by the organization that provides cloud services available to the general public or companies.

A private cloud infrastructure operates exclusively and covers the specific needs of an organization. It is managed by the organization or a third party and may exist on or off the premises.

Nevertheless, there are two other forms of cloud computing infrastructures: the community cloud and the hybrid cloud.

Regarding the community cloud, it is considered a collaborative effort in which infrastructure is shared among several organizations from a specific community with common concerns (compliance, security, etc.), whether managed internally or by a third party and hosted internally or externally.

The hybrid cloud is an infrastructure that employs both private and public cloud services. In this concept, many different types of cloud services are combined in order to meet a variety of end users' (customers') needs.

1.14.2 Cloud GIS Benefits and Risks

Cloud computing together with GIS capabilities are nowadays efficiently blended to create new perspectives and capabilities in GIS applications. The main philosophy in the background of the term *cloud GIS* is seamless access to GIS capabilities from anywhere. Cloud GIS use features, tools, and methods provided by a third-party vendor through Web technologies instead of running GIS software on one server.

It is obvious that this new era for GIS applications (cloud GIS) has significant benefits, the main ones being the following:

Cost Efficiency Before cloud computing, GIS software was an exclusive privilege of companies that could afford high up-front investments in powerful hardware and charged high maintenance costs and software licenses. In contrary to traditional

GIS, cloud GIS provides users with services in a simpler and more affordable way. Also, cloud GIS offers cost-efficient solutions for services and providers, because it charges only for what is used. In cases of open-source solutions, there is no charging.

Software Updates Cloud GIS offers to the end user, in most cases, the latest version of the relevant software or service. There is no need for installing or configuring software updates and one can take advantage of new features right away as they become available.

Publish on the Web While traditional GIS require separate tools for desktop and Web-based operations, as well as extensions and plug-ins, cloud GIS is designed as an all-in-one WebGIS solution. Procedures such as those to create, edit, analyze, and publish data from the GIS service are done in a simple way and the results (like final thematic maps) can automatically be published on the Web. Moreover, in cloud GIS it is very easy to perform data distribution and data sharing. There is no need for sending data using DVDs, using external hard drives, or downloading large datasets and updating local servers.

Collaboration and Synchronization A cloud GIS solution is very useful in cases where large GIS projects are developed or when GIS applications are part of everyday working life. In such cases, a large number of employees across offices, in remote branches, or off-site need to collaborate and synchronize their tasks. In contrast to cloud GIS, in the traditional GIS environment many procedures are implemented exclusively from the users and problems like bugs and capacity limits are difficult to handle.

Access to Datasets For GIS application providers, the cost of obtaining and analyzing datasets from a spatial data vendor is significantly high. A cloud GIS vendor can provide data as a choice of the core services, available through Internet-enabled devices. Cloud GIS provides users with advanced capabilities to input, analyze, and store spatial information.

E-Commerce Cloud GIS provides the capability to sell data or online services very easily and effectively through Web services.

Nevertheless, it is important to mention that there are also some notable risks when working in a cloud architecture. Security and privacy are the two principal concerns. Typical security and privacy issues are dealing with data storage and transfer protection, vulnerability management and remediation, and physical and application security, as well as data privacy. Therefore, it is of great importance to proceed to the choice of a cloud vendor with great experience of such provisions, testimonials, and proven reliability with respect to operating procedures and performance.

In addition, the dependence of cloud GIS on Internet availability and accessibility, although providing many advantages, as already stated, can become disadvantageous if the Internet crashes or when the Internet connection is slow and limited.

1.15 GIS Applications in the Environmental Monitoring

1.15.1 Environment

GIS has a key role in environmental monitoring, modeling, and assessment, as it offers powerful computer mapping and analyses, capable of integrating large volumes of spatial data as well as linking spatial with nonspatial datasets (e.g., census information, environmental exposure levels). For example, observing environmental changes over time indicates trends and patterns via tools for the display and analysis of time series data. This can be achieved by integrating temporal data within GIS. Also, spatial interpolation methodologies can create continuous surface data layers from sample locations and make predictions.

Moreover, GIS is able to represent three-dimensional composites, which are constructed and interactively visualized by creating spatially continuous surfaces or grids, such as geologic structures and/or water-level datasets. Characteristic examples of environmental issues that are handled through GIS include, among many others:

- Land degradation
- Drought
- Forest fires
- Delineation of protected natural ecosystems
- LULC changes
- Air quality
- Marine pollution

1.15.2 Atmosphere, Weather, and Climate

A GIS constitutes a useful tool for scientists and experts in the fields of meteorology, climatology, and atmospheric sciences, helping them to develop more precise weather analysis and numerical model forecasts about weather, climate, and atmospheric composition. GIS can combine different kind of data, such as point measurements from ground stations, satellite imagery, model outputs, radar images, and other scientific instrumentation to provide a realistic approximation of nature and the dynamic evolution of atmospheric motion and physical parameters in time and space (Fig. 1.31).

GIS offer a variety of potential benefits regarding the coupling of surface–atmosphere data/models and efficient means to analyze and visualize atmospheric phenomena. By using not only the mapping capabilities but also the spatial analysis tools and methods of GIS software, many applications are possible, such as integration, visualization, forecasting, and support for decision-making regarding potential natural hazards, climate changes, and atmospheric variations (e.g., a GIS is able to calculate weather impacts on Earth using scientific models, vulnerability studies, dispersal of pollutants, impact assessment, etc.).

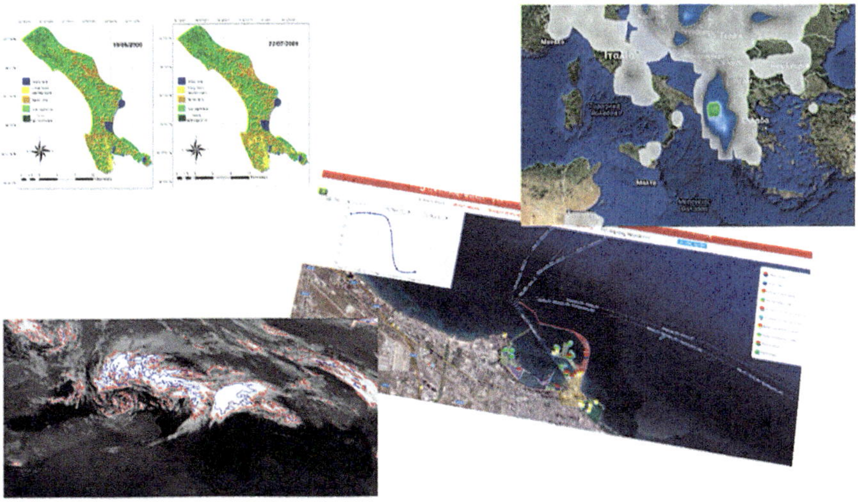

Fig. 1.31 GIS final outputs for different environmental applications

1.15.3 Agriculture

By mapping geographic and geologic features of current (and potential) farmland, scientists and farmers can collaborate in order to develop and implement more effective and efficient cultivating techniques. Such activities could increase food production and improve food quality. With the use of GIS, soil data can be analyzed and combined with well-established farming practices to determine which the best crops to plant are, where they should be cultivated, and how to maintain soil nutrition levels to provide the best possible benefit for the plants (e.g., Fig. 1.32).

GIS applications focused on agriculture not only can map topography and crop health, but can also help in solving wider socioeconomic issues.

GIS capabilities in the agricultural sector can be achieved by integrating high-resolution satellite (or aerial) imagery, field observations, and real-time data and by combining them in order to understand how to make the most of limited resources at any time and any place.

Another important aspect in this field of application is *precision agriculture* (PA)—also called *precision farming* (PF)—which utilizes geographical information to determine field variability and to ensure optimal use of inputs and maximize the output from a farm. Therefore, the combined use of satellite remote sensing, GIS, and GPS can help farmers to more effectively use expensive resources, such as fertilizers, pesticides, and herbicides, and to efficiently use water resources. This modern approach in agriculture not only maximizes the crop yields, but also reduces the operating costs and consequently increases the farmers' profits.

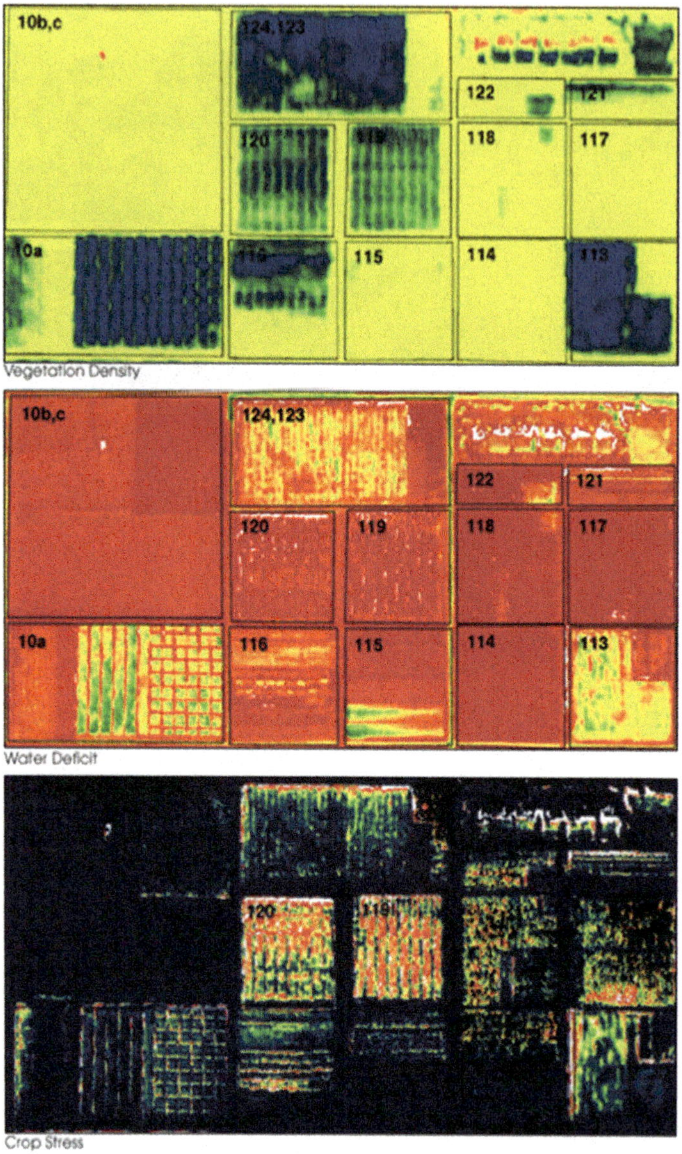

Fig. 1.32 False-color images demonstrating remote sensing applications in precision farming (Courtesy of NASA Earth Observatory)

References

Armstrong L, Butler K, Settlemaier J, Vance T, Wilhelmi O (2015) Mapping and modeling weather and climate with GIS. ESRI Press, New York

Ball JE, Luk KC (1998) Modeling spatial variability of rainfall over a catchment. J Hydrol Eng 3:122–130

Baumann P (2012) GML application schema—coverages. Open Geospatial Consortium. Available at https://portal.opengeospatial.org/files/?artifact_id=48553

Berry R, Higgs G, Fry R, Langford M (2011) Web-based GIS approaches to enhance public participation in wind farm planning. Trans GIS 15(2):147–172

Bhatt G, Kumar M, Duffy JC (2014) A tightly coupled GIS and is distributed hydrologic modeling framework. Environ Model Softw 62:1–15

Brown I, Jude S, Koukoulas S, Walkden M (2006) Dynamic simulation and visualization of coastal erosion. Comput Environ Urban Syst 30(6):840–860

Burrough PA (1986) Principles of geographical information systems for land resources assessment. Oxford University Press, Oxford

Buschmann F, Meunier R, Rohnert H, Sommerlad P, Stal M (2001) Pattern-oriented software architecture: a system of patterns, vol 1, Wiley, New Delhi. ISBN: 978-0-471-95869-7

Clarke KC (1986) Advances in geographic information systems. Comput Environ Urban Syst 10:175–184

De Kluijver H, Stoter J (2003) Noise mapping and GIS: optimising quality and efficiency of noise effect studies. Comput Environ Urban Syst 27(1):85–102

Frehner M, Brandli M (2006) Virtual database: spatial analysis in a web-based data management system for distributed ecological data. Environ Model Softw 21(11):1544–1554

Fu P, Sun J (2010) WebGIS: principles and applications. ESRI Press. ISBN: 9781589482456

Goodchild MF (1985) Geographic information systems in undergraduate geography: a contemporary dilemma. Oper Geogr 8:34–38

Goodchild MF, Proctor J (1997) Scale in a digital geographic world. Geogr Environ Model 1:5–23

Gounaridis D, Zaimes G, Koukoulas S (2014) Quantifying spatio-temporal patterns of forest fragmentation in Hymettus Mountain, Greece. Comput Environ Urban Syst. doi:10.1016/j.compenvurbsys.2014.04.003

Grance T, Mell P (2011) Computer security. National Institute of Standards and Technology, US Department of Commerce, Special Publication 800–145. http://nvlpubs.nist.gov/nistpubs/Legacy/SP/nistspecialpublication800-145.pdf

Hill J, Stellmes M, Udelhoven T, Sommer S (2008) Mediterranean desertification and land degradation. Mapping related land use change syndromes based on satellite observation. Glob Planet Chang 64(3–4):146–157

Huang Q, Yang C, Nebert D, Liu K, Wu H (2010) Cloud computing for geosciences: deployment of GEOSS Clearinghouse on Amazon's EC2, HPDGIS '10. In: Proceedings of the ACM SIGSPATIAL International Workshop on High Performance and Distributed Geographic Information Systems, pp 35–38

Johnson S (2006) The story of London's most terrifying epidemic—and how it changed science, cities, and the modern world. Riverhead Books, New York

Julien Y, Sobrino JA, Jiménez-Muñoz J-C (2011) Land use classification from multitemporal Landsat imagery using the yearly land cover dynamics (YLCD) method. Int J Appl Earth Obs Geoinf 13:711–720

Kolios S, Stylios C (2013) Identification of land cover/land use changes in the greater area of the Preveza peninsula in Greece using Landsat satellite data. Appl Geogr 40:150–160

Kolios S, Stylios C, Petunin A (2015) AWebGIS platform to monitor environmental conditions in ports and their surroundings in South Eastern Europe. Environ Monit Assess. doi:10.1007/s10661-015-4786-x

Koukoulas S, Blackburn GA (2005) Mapping individual tree location, height and species in broadleaved deciduous forest using airborne LIDAR and multi-spectral remotely sensed data. Int J Remote Sens 26(3):431–455

Koulouri M, Giourga C (2007) Land abandonment and slope gradient as key factors of soil erosion in Mediterranean terraced lands. Catena 69(3):274–281

Kraak M-J, Brown A (2001) Web cartography. In: Kraak MJ, Brown A (eds) Web cartography. Taylor and Francis, New York

Krige DG (1951) A statistical approach to some basic mine valuation problems on the Witwatersrand. J Chem Metall Min Soc S Afr 52(6):119–139

Lee Y, Chen K (2010) Is server consolidation beneficial to MMORPG? A case study of World of Warcraft. In: Proceedings—2010 I.E. 3rd international conference on cloud computing, CLOUD 2010, 5–10 July 2010, pp 435–442

Liying G, Daolong W, Jianjun Q, Ligang W, Yiu L (2009) Spatio-temporal patterns of land use change along the Bohai Rim in China during 1985–2005. J Geogr Sci 19:568–576

Longley A, Goodchild F, Maguire J, Rhind J (2010) Geographical information systems and science, 3rd edn. Wiley, Chichester

Louca M, Vogiatzakis I, Moustakas A (2015) Modelling the combined effects of land use and climatic changes: coupling bioclimatic modelling with Markov-chain cellular automata in a case study in Cyprus. Eco Inform. doi:10.1016/j.ecoinf.2015.05.008

Maguire DJ, Goodchild MF, Batty M (2005) GIS, spatial analysis and modeling. ESRI Press, New York

Maliene V, Grigonis V, Palevičius V, Griffiths S (2011) Geographic information system: old principles with new capabilities. Urb Des Int 16(1):1–6. doi:10.1057/udi.2010.25

Matheron G (1963) Principles of geostatistics. Econ Geol 58(1246 1266):1963

McHarg LI (1992) Design with nature: 25th anniversary edition. Wiley, Canada

Mell P, Grance T (2009) The NIST definition of cloud computing, Ver.15. Available at: http://csrc.nist.gov/groups/SNS/cloud-computing/

Mumford I (1999) Milestones in lithographed cartography from 1800s. PhD thesis, University of Reading

Muzafar Ahmad Bhat, Razeef Mohd Shah, Bashir Ahmad (2011) Cloud computing: a solution to geographical information systems (GIS). Int J Comput Sci Eng 3(2):594–600, ISSN:0975-3397

Olson AJ (2010) Data as a service: are we in the clouds? J Map Geogr Libr Adv Geospat Inf Collect Arch 6(1):76–78

Papakonstantinou A, Varsamis D, Soulakellis N (2011) Inset Mapper: a software tool in island cartography. Cartogr Geogr Inf Sci 38(4):384–397

Papakonstantinou A, Kontos T, Christodoulou E, Soulakellis N (2013) 3D geovisualisation of noise and visual impact of a proposed wind farm development using a GIS based visual–acoustic 3D simulation. In: 26th international cartographic conference held in Dresden in August 2013

Pascual M, Alves E, De Almeida T, Holanda M (2012) An architecture for geographic information systems on the Web—WebGIS. The fourth international conference on advanced geographic information systems, applications, and services, pp 209–2014. ISBN:978-1-61208-178-6

Petropoulos PG, Kontoes C, Keramitsoglou I (2011) Burnt area delineation from a uni-temporal perspective based on Landsat TM imagery classification using support vector machines. Int J Appl Earth Obs Geoinf 13:70–80

Rigaux P, Scholl M, Voisard A (2002) Spatial databases with application to GIS. Elsevier Science, San Francisco

Seine (1971) Commission instituée pour recueillir les faits relatifs à l'invasion et aux effets du choléra dans le département de la Seine, 1971. Rapport sur la marche et les effets du choléra-morbus dans Paris et les communes rurales du département de la Seine [Microforme]: année 1832/par la Commission nommée, avec l'approbation de M. le ministre du Commerce et des travaux publics, par MM. les préfets de la Seine et de police. Paris: Hachette, http://gallica.bnf.fr/ark:/12148/bpt6k842918/f353.image

Symeonakis E, Caccetta P, Koukoulas S, Furby S, Karathanasis N (2011) Multi-temporal land-cover classification and change analysis with conditional probability networks: the case of Lesvos Island (Greece). Int J Remote Sens. doi:10.1080/01431161.2011.640961

Symeonakis E, Karathanasis N, Koukoulas S, Panagopoulos G (2014) Monitoring sensitivity to land degradation and desertification with the Environmentally Sensitive Area Index: the case of Lesvos Island. Land Degrad Dev 2014. doi:10.1002/ldr.2285

Tomlinson RF (1967) An introduction to the geo-information system of the Canada Land Inventory. Canada Department of Forestry and Rural Development, Ottawa

Tzanopoulos J, Vogiatzakis I (2010) Processes and patterns of landscape change on a small Aegean island: the case of Sifnos, Grecce. Landsc Urban Plan 99(1):58–64

Yang C, Goodchild M, Huang Q, Nebert D, Raskin R, Xu Y, Bambacus M, Fay D (2011) Spatial cloud computing: how geospatial sciences could use and help to shape cloud computing? Int J Digit Earth 4(4):305–329

Yuan F, Sawaya EK, Loeffelholz CB, Bauer EM (2005) Land cover classification and change analysis of the Twin Cities (Minnesota) metropolitan area by multitemporal Landsat remote sensing. Remote Sens Environ 98:317–328

Chapter 2
WebGIS Applications for Weather, Marine, and Atmospheric Environments

During the last decade, there has been an increasing tendency for development of GIS applications regarding the environment. These applications not only focus on spatial analysis, decision making and future planning using the wide range of GIS capabilities, but they also provide to end users and the public integrated Web-based applications visualizing the final results of scientific efforts via maps and graphs (e.g., Fustes et al. 2014; Kulkarni et al. 2014). These computer-based applications are essential for modern visual representation of environmental issues through the monitoring of spatial and temporal changes, even in remote areas, helping planners and stakeholders to design and envision, on medium- to long-term scales, an environmentally sustainable future.

In this chapter, four characteristic examples/case studies of Web-based GIS applications developed mainly by members of the Laboratory of Knowledge and Intelligent Computing of the Technological Educational Institute of Epirus (Greece) are presented: the TEN ECOPORT WebGIS platform for marine environmental monitoring (www.tenecoport.eu/webgis), the PORTWEATHER WebGIS application for port weather monitoring/forecasting (www.portweather.eu), the SevereWeather application for extreme weather events (http://qgiscloud.com/ko_sta/Weather_extreme_events_Greece) and, finally, the SatWeather application for satellite weather monitoring/forecasting (http://satweather.kic.teiep.gr).

2.1 WebGIS Application for Marine Environmental Monitoring

2.1.1 Introduction

The continuous increase of international trade, the new tendencies in transportation (using modern, larger, and safer ships) and the advances in technology, highlight

S. Kolios et al., *GIS and Environmental Monitoring*, Geotechnologies and the Environment 20, DOI 10.1007/978-3-319-53086-4_2

ports as major factors in the development of cities, regions, and countries (Ng and Liu 2014).

However, port-related activities and the explosion of maritime commerce through intermodal and unimodal transportation cause serious impacts on the quality of the marine and air environments (e.g., Dinwoodie et al. 2012; Grifoll et al. 2011; Gupta et al. 2005; Bailey and Solomon 2004; Edoho 2008; Eyring et al. 2010). Such activities have many significant consequences for the health and overall quality of life of the populations that live close to ports, as well as for the natural environment (Eyring et al. 2010; WHO 2000). Ports attract a constantly increasing amount of ship traffic, thus influencing the ecosystems of the nearby port areas. Such ecosystems are considered highly vulnerable areas (Urquhart et al. 2013) and include estuaries, surrounding coastal waters, and protected natural zones [such as Sites of (European) Community Importance (SCI) and Special Protected Areas (SPA) in terms of their biodiversity].

Nevertheless, there is a two-way interaction between seaports and the natural environment. The port areas, maritime shipping, and sea commerce are highly sensitive to extreme weather phenomena, causing serious impacts on any port-related activities (Becker et al. 2012; Nursey–Bray et al. 2013).

Nowadays, environmental issues play a crucial role in the majority of port activities, the logistics chain, and management plans and practices. The sustainability of port development and management affect port authorities and their connection with port stakeholders and vice versa. Also, port authorities are increasingly aware of costs in terms of financial penalties for neglecting their environmental duties and for failing to apply effective environmental strategies. The environmental policy usually comprises environmental management strategies, marine safety, coastal zone protection, conservation of the flora and fauna of the local ecosystems, air and marine pollution, and waste and infrastructure development projects. In this aspect, WebGIS applications are considered to be integral parts of management action plans (MAPs) and improvement of seaport environmental policies (Marinski et al. 2015).

Moreover, a well-known and widely applied practice that contributes nowadays to the environmentally sustainable perspectives of ports that are operating under highly competitive conditions is the development of environmental management systems (EMS), the use of environmental performance indicators (EPI), and the implementation of international common practices defined by audit schemes (e.g., Puig et al. 2014; Donnelly et al. 2007; Quynh et al. 2011; PPRISM 2012; Portman 2014).

To this end, the role of satellite data products is crucial for monitoring and improving many port-related activities, providing continuous and accurate information regarding the environment. More specifically, many modern satellite missions are solely focused on the collection of datasets using instrumentation on board while the relevant satellite platforms are able nowadays to provide accurate remotely sensed data and products covering wider areas that it is impossible to continuously monitor using other instrumentation or technologies.

2.1.2 The TEN ECOPORT WebGIS Platform

Within this context and taking advantage of all the available data and tools that provide useful outcomes about environmental issues in the marine environment and port areas (e.g., Fustes et al. 2014; Kulkarni et al. 2014; Noyon and Devogele 2005), the TEN ECOPORT WebGIS platform (www.tenecoport.eu/webgis) was developed. It is a modern Web-based GIS application that gathers and illustrates useful information from satellite datasets, as well as measurements from environmental stations operating inside the port areas of a network of 12 ports in South East Europe, i.e., the ports of Bari, Barletta, Monopoli, and Brindisi (Italy), Igoumenitsa and Patras (Greece), Burgas and Varna (Bulgaria), Constanta (Romania), Bar (Montenegro), Dubrovnik (Croatia), and Durres (Albania).

The TEN ECOPORT WebGIS platform was developed within the framework of the Transnational ENhancement of ECOPORT8 network (TEN ECOPORT) project (http://www.tenecoport.eu), cofunded by the Transnational Cooperation Programme South East Europe (SEE programme). It provides valuable information about environmental conditions and possible sources of pollution, as well as spatial and temporal distributions of values for several parameters concerning seawater quality with the use of point measurements in port stations and satellite data products.

The application was designed based on the needs of the TEN ECOPORT partnership, but with the scope to cover any potential and future needs of a wider community of users, such as port authorities, universities, researchers, and anyone interested in the environmental conditions in South East Europe (SEE) ports and their commercial sea corridors. The central interface of the TEN ECOPORT WebGIS platform can be seen in Fig. 2.1.

The portability of the WebGIS platform (easily accessible via mobile devices, such as tablets and smartphones) makes this platform not only very informative regarding environmental issues of SEE ports and their sea corridors, but also easy to use.

It is noteworthy that for the design and the development of this Web-based application, the four general monitoring guidelines proposed in the study of Ferreira et al. (2007) (i.e., definition of appropriate objectives; setting priorities and optimization; implementation of quality control; assessment of monitoring success) were followed. Initially, the issues and the potential needs were defined, after official meetings and analytic discussion with all involved partners (port authorities, institutions, and experts) that expressed the intention to provide information within the framework of the TEN ECOPORT project (http://www.tenecoport.eu). Then, the optimum solutions, not only from a technical but also from an environmental point of view, were set. After recording the needs, all the appropriate datasets and the relevant informational material were selected, while internal quality controls were applied where necessary (e.g., Kolios and Stylios 2015). The assessment of the application followed the development of the TEN ECOPORT WebGIS platform.

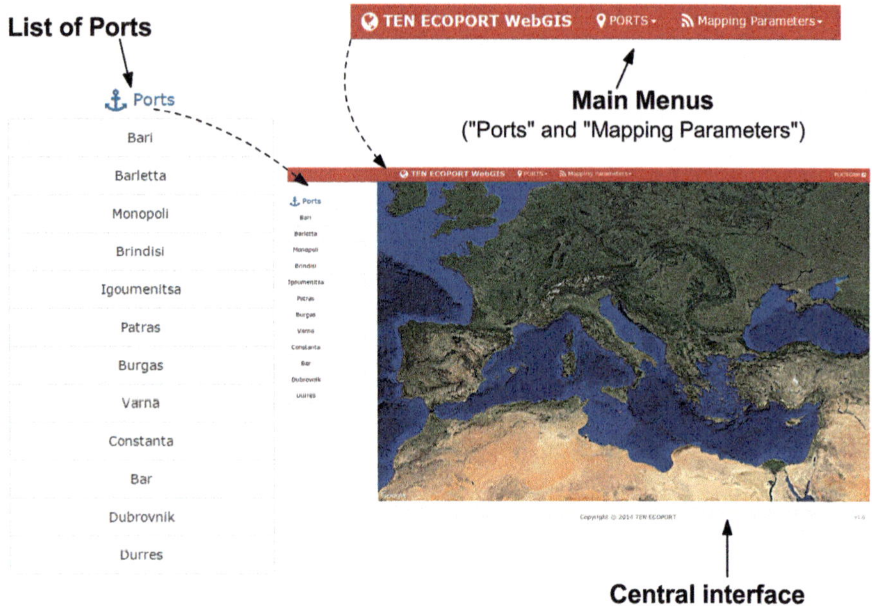

Fig. 2.1 Central interface of the TEN ECOPORT WebGIS platform (http://www.tenecoport.eu/webgis) providing access to the two main menus and the list of available ports

During this stage, the functionality, the portability, and the coverage of the initially defined needs were checked during meetings with all partners.

2.1.3 *Environmental Issues in Port Areas*

In order to develop the application that would provide useful data and information on the environmental conditions in port areas and their sea corridors, feedback from port environmental offices, port authorities, and other relevant stakeholders was necessary. The environmental issues regarding air and water in the abovementioned 12 ports were recorded. It is also worth pointing out that particle matter and gas emissions, as well as waste management (in water and soil), were included among the most important environmental issues of the ports examined in the framework of the TEN ECOPORT project.

2.1.4 *Datasets*

The TEN ECOPORT WebGIS platform includes several kinds of data. More specifically, satellite-derived data and point measurements from stations in port

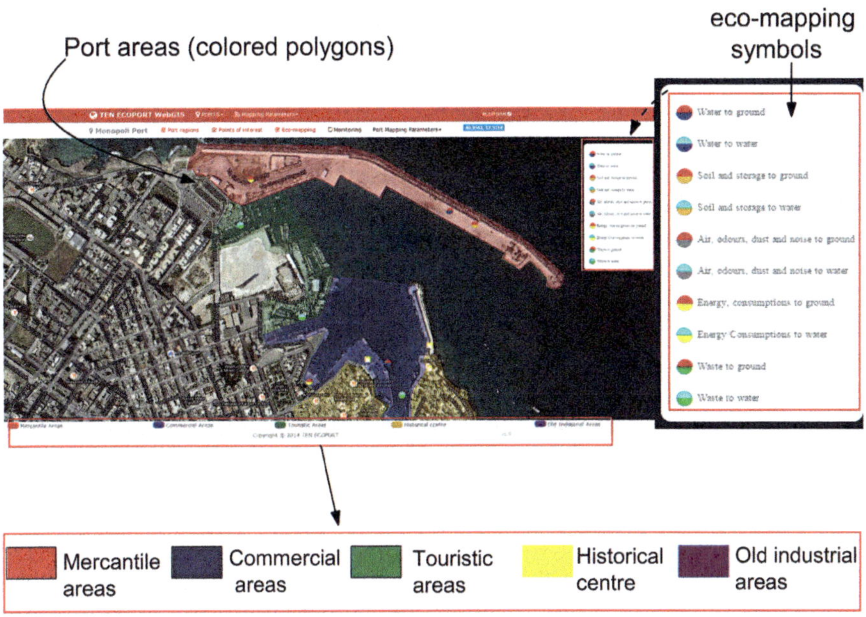

Fig. 2.2 Monopoli port area, when the "Point of interest" and "Eco-mapping" submenus in the "PORTS" menu of the WebGIS platform are selected. Different sectors inside port areas are depicted with *polygons*, and environmentally sensitive locations are *highlighted* by *points in different colors*

areas were collected and used. Information provided by the environmental offices of port authorities with regard to the locations and the areas of specific environmental interest was also used in order to place suitable symbols at the right points so as to highlight points and areas with environmental issues (Fig. 2.2).

The main purpose regarding the use of satellite products is to highlight short-term mean spatial changes of basic sea parameters on a monthly basis. This approach is intended to give a clear and accurate depiction of areas that are environmentally sensitive, aiming to operate as a valuable informational background for any interested agency, organization, or company (at the local, national, and European levels) that may need to detect spatial patterns of sea surface temperature (SST) and chlorophyll *a* values. This visual representation supports decision making, proposing solutions, and adapting new strategies or legislation for a sustainable future of the sea environment and for the protection of coastal areas, especially around ports.

More specifically, three basic parameters were chosen to cover the abovementioned aspects: SST (in degrees Celsius), chlorophyll *a* (in milligrams per cubic meter) and colored (also called chromophoric) dissolved organic matter (CDOM) (unitless). These parameters are key factors for the water quality and such information can be very useful especially for geographical areas such as South East Europe, which includes many areas with eutrophic problems and which is affected by—among other things—many human activities, sea commerce, and tourist activities (e.g., Karydis and Kitsiou 2012).

Before analytically presenting the selected parameters, it must be mentioned that the original satellite datasets came from NASA's MODIS (MODerate resolution Imaging Spectroradiometer) instrument on board the Terra and Aqua satellite platforms, and they are widely used in a great variety of studies (e.g., Shang et al. 2011; Urquhart et al. 2013; Li et al. 2013; Brewin et al. 2013).

The data regarding SST, which represents the temperature of the water at the topmost layer of the water surface, were selected on a monthly basis for a period between 2000 and 2013 (a total of 168 files) in a spatial resolution of 9 km. The data products of chlorophyll *a* and CDOM were selected on a monthly basis for a period between 2003 and 2013 (a total of 132 files) at a spatial resolution of 4 km.

Chlorophyll *a* is a data product generated by the NASA Ocean Biogeochemical Model (NOBM) based on data assimilation of remotely sensed chlorophyll *a*. This type of chlorophyll is the most common and predominant in all oxygen-evolving photosynthetic organisms and it is strictly connected with the phenomenon of eutrophication.

CDOM absorbs light, most strongly short-wavelength light ranging from blue to ultraviolet. CDOM diminishes light as it penetrates water, thus inhibiting the growth of phytoplankton populations, which form the basis of sea food chains and are a primary source of atmospheric oxygen. CDOM also absorbs harmful UVA/B radiation, protecting organisms from DNA damage.

At this point, it must be mentioned that as we had available datasets of point measurements from ground stations about SST and chlorophyll *a*, we evaluated, in selected samples and ports, the quality of the satellite datasets by comparing them with the relevant point measurements and we concluded that the accuracy of the monthly satellite datasets was satisfactory (Kolios and Stylios 2015).

For each of the three selected parameters, we calculated mean monthly values (per pixel) for the whole period of the relevant datasets. Figure 2.3 provides a schematic representation regarding the calculation of the mean values (per pixel). Finally, the mean values per pixel were used to create mapping representations for the spatial distributions of the mean values for every parameter.

Referring to the point measurements, local stations that measure environmental parameters of the sea and/or air quality operate in seven of the 12 ports collaborating within the framework of the TEN ECOPORT project. Many of the recorded measurements were provided and presented through the TEN ECOPORT WebGIS platform. For each of these parameters, a menu was developed to plot their temporal evolution. This menu may include possible future updates in order to provide relevant graphs on a real-time basis. In its current version, the TEN ECOPORT WebGIS platform provides graphs for sample datasets (archive data) on the parameters and the ports referred to in Table 2.1.

A schematic diagram of the data and the information flows in the TEN ECOPORT WebGIS platform are given in Fig. 2.4.

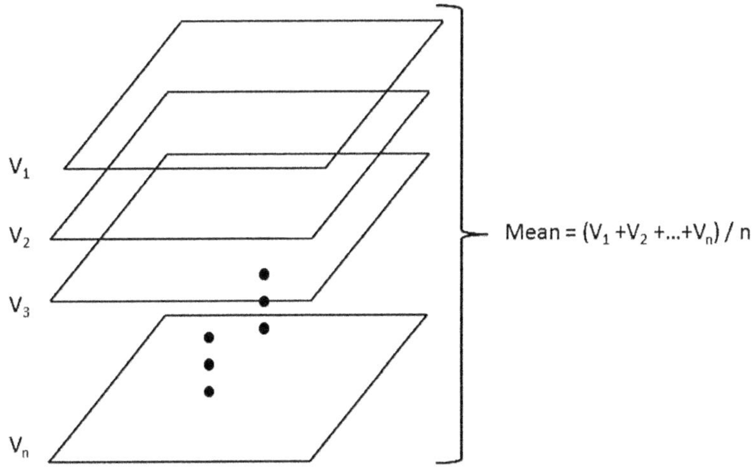

Fig. 2.3 Schematic representation regarding the calculation of the mean values (per pixel) (Kolios et al. 2015)

Table 2.1 List of measured parameters displayed on the TEN ECOPORT WebGIS platform, as provided by the environmental stations installed in ports of the SEE ports network

n/n	Station location	Country	Parameters
1	Port of Bari	Italy	PM_1, $PM_{2.5}$, PM_4, PM_{10}, PST
2	Port of Barletta	Italy	PM_1, $PM_{2.5}$, PM_4, PM_{10}, PST
3	Port of Igoumenitsa	Greece	CO, NO_2, SO_2, O_3
4	Port of Patras	Greece	PM_1, $PM_{2.5}$, PM_{10}
5	Port of Burgas	Bulgaria	Turbidity, PH, salinity, chlorophyll, temperature
6	Port of Constanta	Romania	Wind (direction and speed), relative humidity, air temperature
7	Port of Bar	Montenegro	SST, BOD, Salinity, chlorophyll *a*

2.1.5 Main Menus

Two main menus are provided through the TEN ECOPORT WebGIS platform central interface: the "PORTS" menu and the "Mapping Parameters" menu. A schematic diagram of these menus as well as their submenus is given in Fig. 2.5.

The "PORTS" Menu and Its Submenus

The main menu of the TEN ECOPORT WebGIS platform, named "PORTS," provides information mainly focusing on port stakeholders, port authorities,

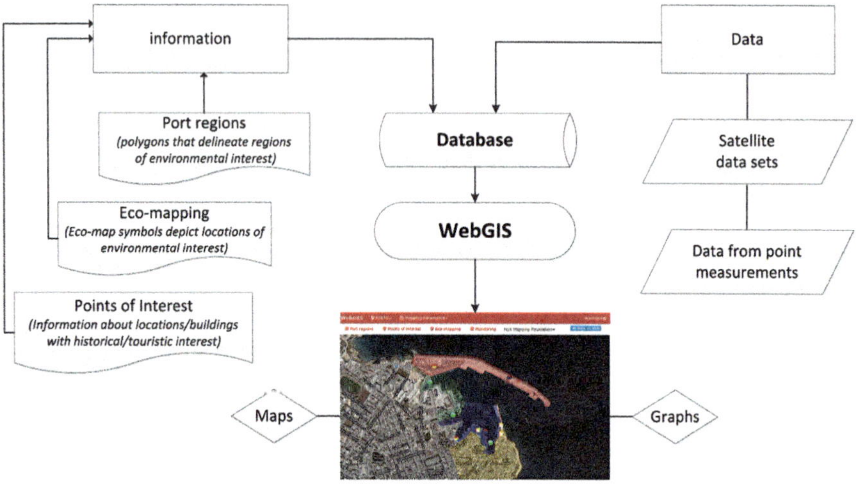

Fig. 2.4 Diagram of the data and the information flows provided through the TEN ECOPORT WebGIS platform (Kolios et al. 2015)

touristic and transportation stakeholders, and the general public. The user can choose the port of interest and then one (or more) of the following submenus: "Port regions," "Points of interest," "Eco-mapping," "Monitoring," and "Port Mapping Parameters."

The "Port Regions" Menu By selecting the Port regions submenu, the main areas of interest in a specific port territory appear. Colored polygons have been designed to represent specific locations of ports. Each one of the selected colors, i.e., red, blue, green, yellow, and magenta, is dedicated to representation of one specific area of interest, i.e., Mercantile areas, Commercial areas, Touristic areas, Historical centre, and Old Industrial areas, respectively (Fig. 2.6).

The "Eco-mapping" Submenu Another important informational field of the TEN ECOPORT WebGIS platform is included in the "Eco-mapping" submenu. This submenu highlights environmentally sensitive locations or locations of high interest in the port area. More precisely, the "Eco-mapping" tool was designed and introduced as an original and simple tool, which helps companies in the implementation of environmental management through symbols that depict locations of high environmental importance inside ports. The "Eco-mapping" tool constitutes a first step toward port environmental management and aims to be not only an inventory of good practices and interest points, but also a systematic method of conducting on-site environmental review and a tool that encourages stakeholders' involvement and participation.

The "Eco-mapping" symbols were designed to cover the main environmental issues at ports, gathering all the aspects that are related to water, energy, soil, waste management, etc. Figure 2.7 presents the "Eco-mapping" symbols along with their meanings. The color of the upper part of the circle indicates the source of the

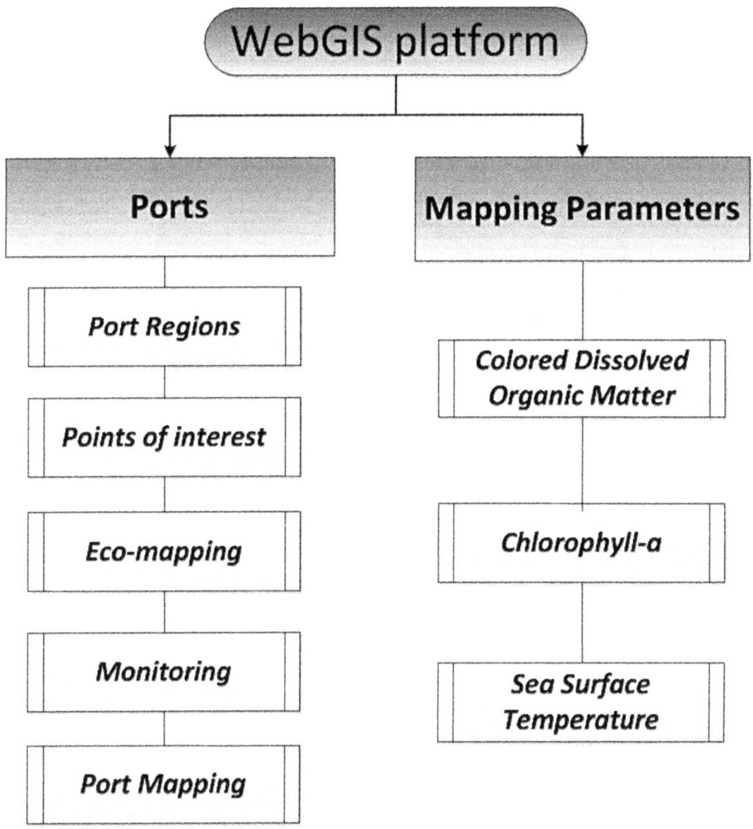

Fig. 2.5 Main menus ("PORTS" and "Mapping Parameters") and submenus of the TEN ECOPORT WebGIS platform (Kolios et al. 2015)

possible pollution and the color of the lower part of the circle indicates the issue. More specifically, for the upper part of the circle, the red color shows that the pollution comes from the ground and the light blue (cyan) shows pollution of water. Regarding the colors for the lower part of the circle, the blue color represents water surfaces; the brown is for soil and storage; the gray color is for air, odors, dust, and noise; the yellow color is for energy/consumption; and the green represents waste. For example, the circle with the red-colored upper part and the blue-colored lower part (the circle in the upper left part of Fig. 2.7) must be placed in locations in sea areas where there are discharges from the land (Marinski et al. 2015).

Therefore, the relevant "Eco-mapping" symbols have been placed in every port to highlight critical environmental issues, as shown in the screenshot examples in Fig. 2.8. It should be mentioned that when a user clicks on a symbol, detailed environment-related information about the specific location appears.

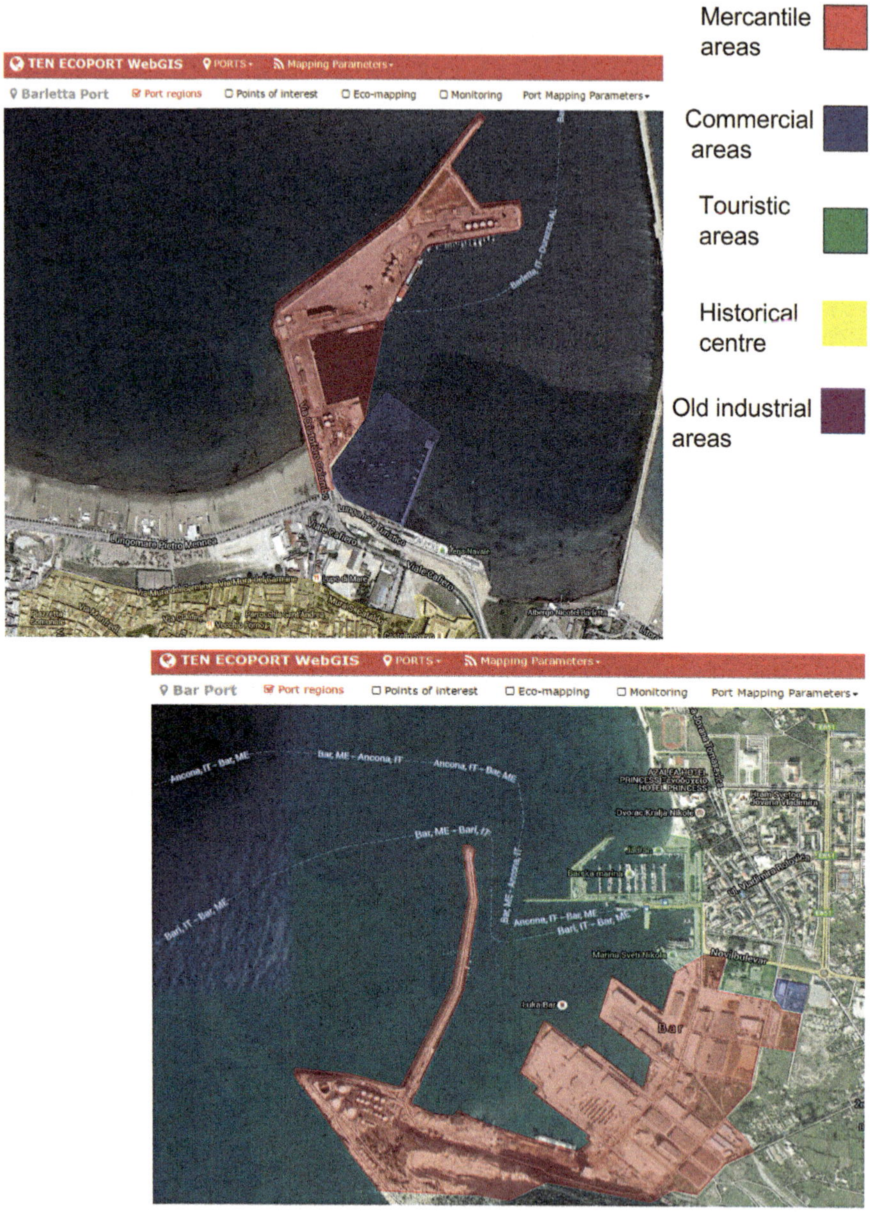

Fig. 2.6 Two examples from the ports of Bar and Barletta, where *differently colored polygons* depict port areas that contribute to (or are affected by) the environmental conditions of the wider port area

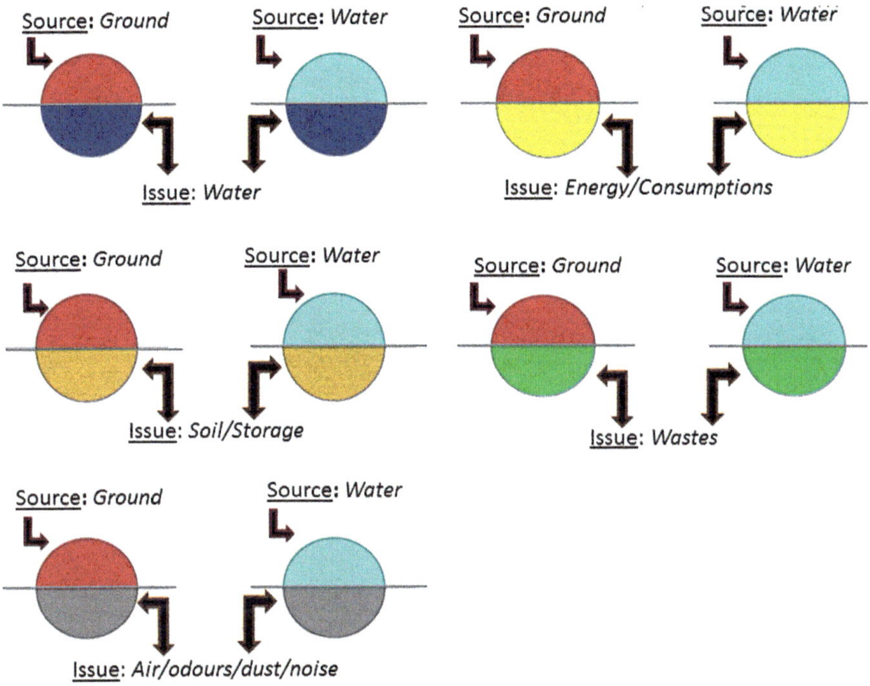

Fig. 2.7 "Eco-mapping" environmental symbols and their meanings. The *upper parts of the circles* indicate the source of pollution and the *lower parts* indicate the issue

The "Points of Interest" Submenu Besides the circular environmental ("Eco-mapping") symbols, rectangular yellow symbols that highlight buildings and infrastructure of historical interest within a specific port area appear when the "Points of interest" submenu is selected. This information is considered to be of significant importance for stakeholders and constitutes a first effort to record touristic/historical information around port areas. The coexistence of information regarding buildings and infrastructure of historical interest together with the information provided as the result of the "Eco-mapping" could help to plan and design actions and policies to protect monuments and buildings of historical importance from nearby sources of environmental pollution (Fig. 2.9).

The "Monitoring" Submenu This submenu provides the user, via a popup window, with the capability to create and present his own graph by choosing the preferred parameter and period of interest using the three available drop-down lists (Fig. 2.10). The list of parameters concerns both atmospheric and marine quality parameters (e.g., particle matter, salinity, chlorophyll, CO). The user can choose the parameter and the period of interest, and then the temporal evolution of the measurements is provided through the relevant graph.

The "Port Mapping Parameters" Submenu The mean monthly spatial distributions of the three basic parameters (CDOM, SST and chlorophyll *a*) in the greater

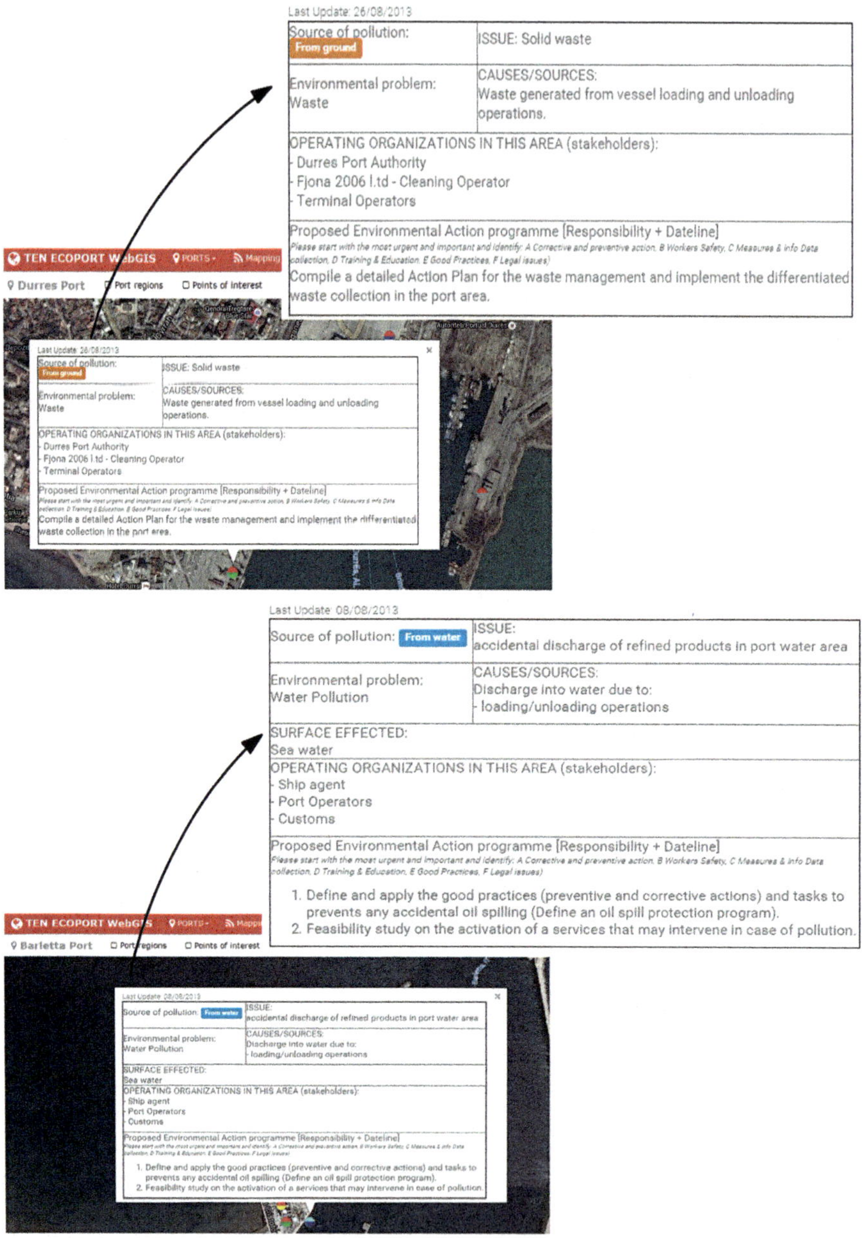

Fig. 2.8 Visualization of two port areas (the port of Durres and the port of Barletta) when the "Eco-mapping" submenu in the "PORTS" menu of the TEN ECOPORT WebGIS platform is selected

Fig. 2.9 Information regarding monuments and buildings of historical importance located near sources of environmental pollution, provided through the TEN ECOPORT WebGIS platform

area of the selected port is available through the "Port Mapping Parameters" submenu. The user chooses the parameter and the month of interest, and the mean monthly spatial distribution of the corresponding parameter is provided (Fig. 2.11). It should be mentioned that the smooth color changes are due to the spatial resolution of the satellite data (4–9 km). This means that there are not abrupt changes in values for small sea areas and the color changes are smoothed for better visualization. Nevertheless, the actual pixel values are not affected. If the user clicks on the map, the actual values are provided as well as the distance from the nearest pixel center (the percentage value below the parameter value).

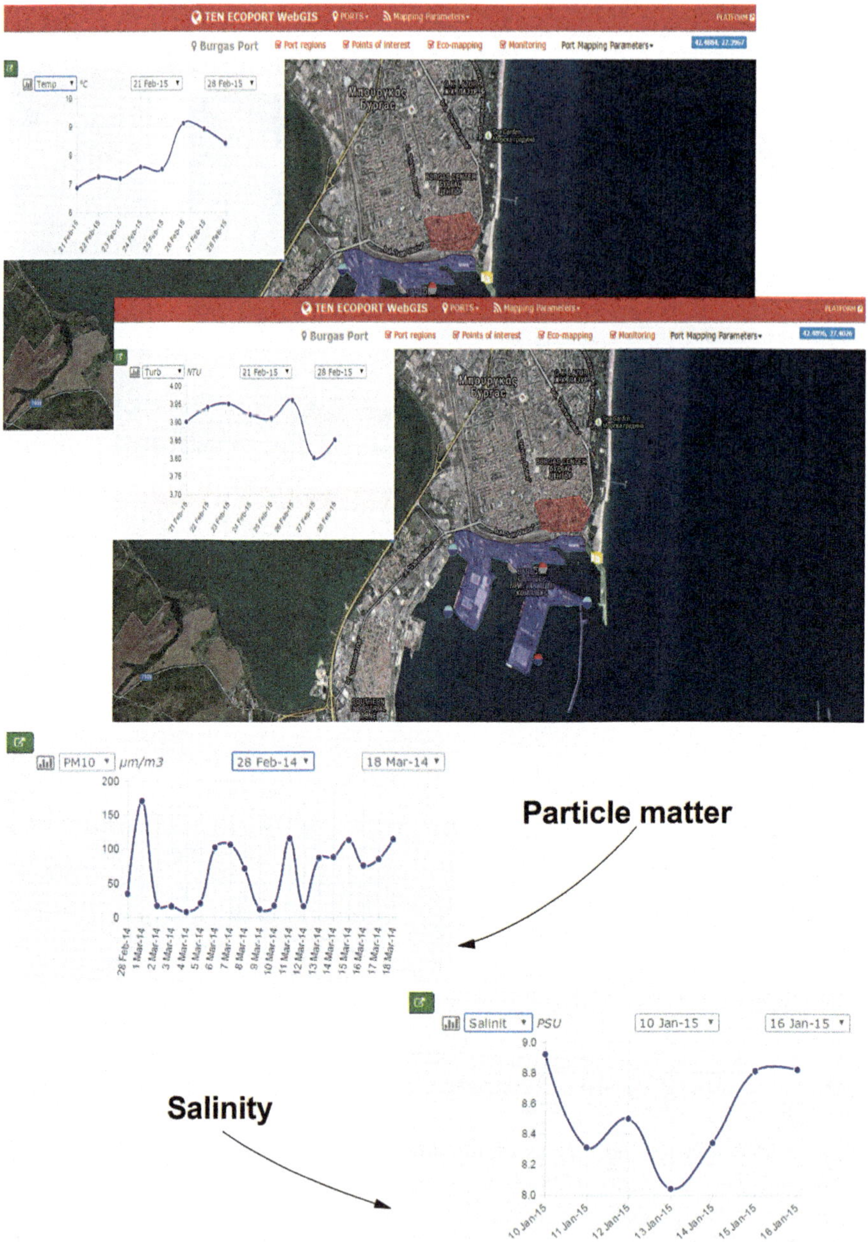

Fig. 2.10 "Monitoring" submenu in the "PORTS" menu of the TEN ECOPORT WebGIS platform

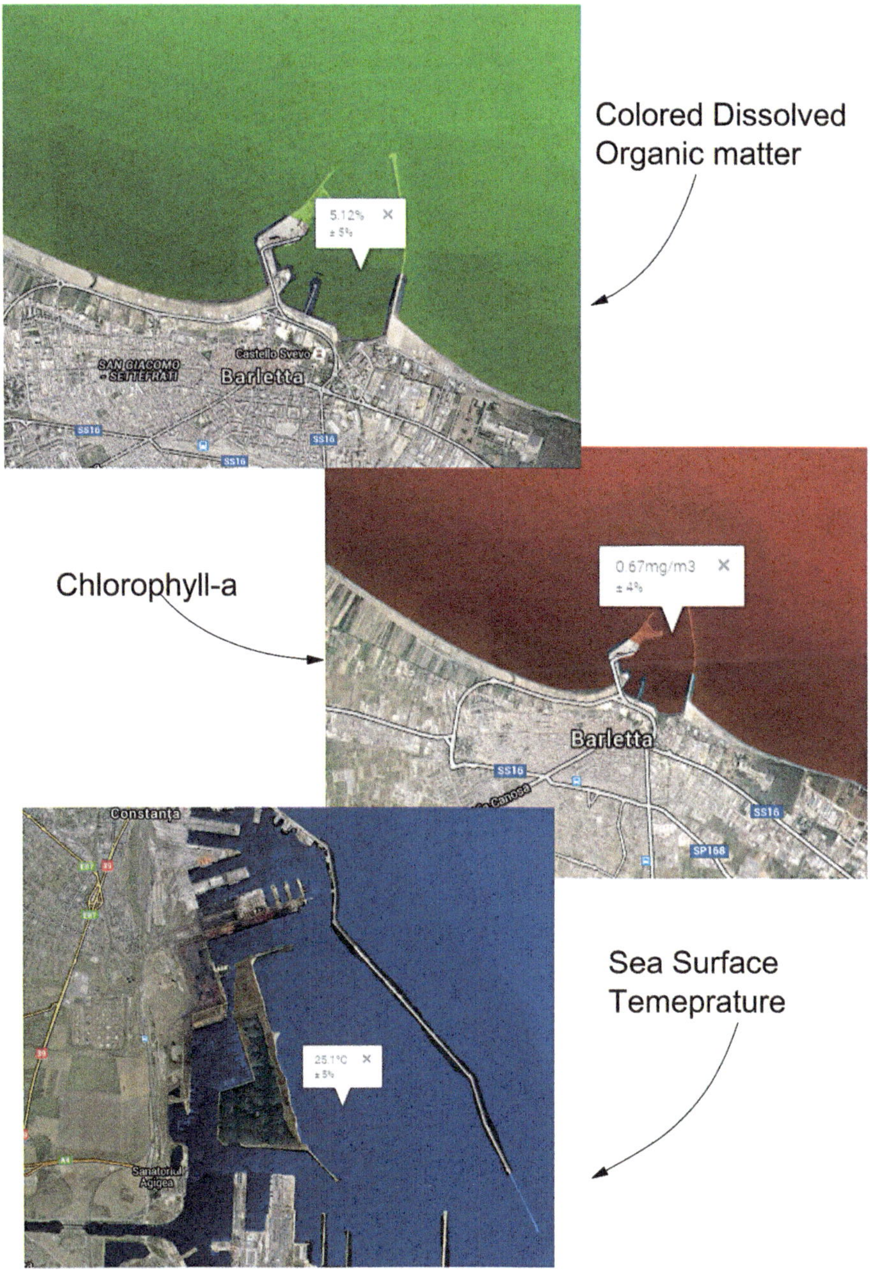

Fig. 2.11 Examples of the spatial distributions of the monthly mean values of the parameters of colored dissolved organic matter, chlorophyll *a*, and sea surface temperature

The "Mapping Parameters" Menu

Another environmentally important menu under the title "Mapping Parameters" reveals interesting information addressed mainly to the scientific community and policy makers. This menu hosts maps of the spatial distributions of monthly satellite-derived products (SST, chlorophyll *a* and CDOM) for the whole of South East Europe. Examples of the visualization of these parameters are given in Fig. 2.12.

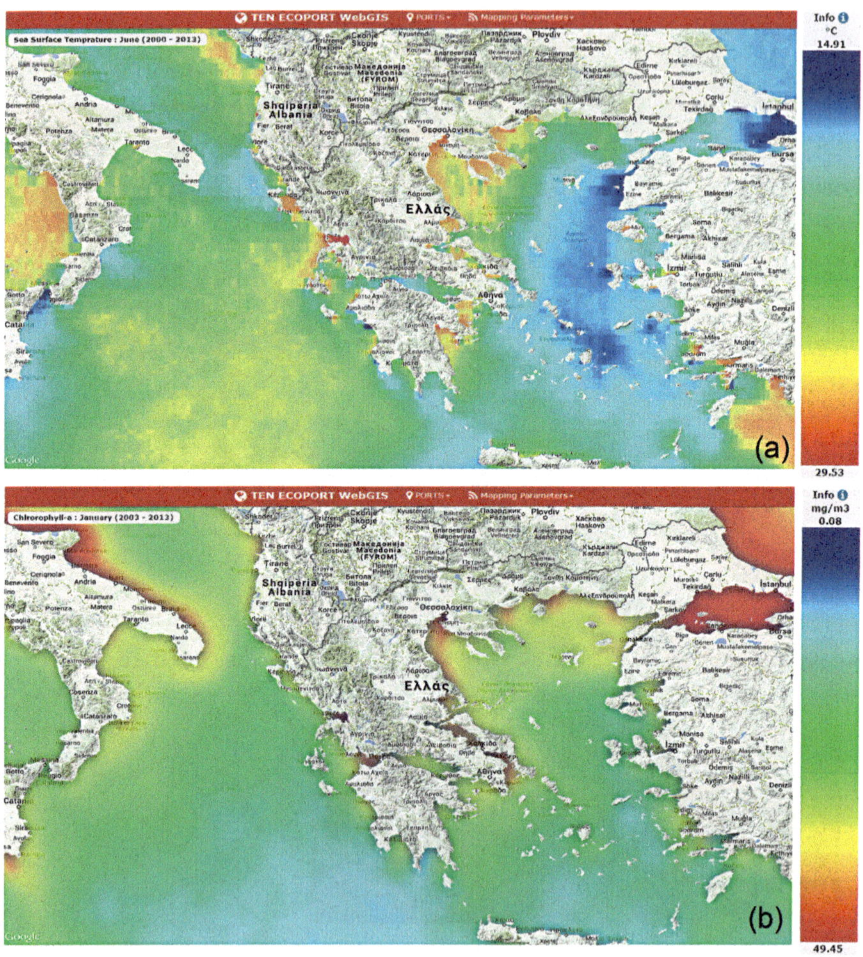

Fig. 2.12 (**a**) Example of the spatial distribution of the mean monthly (June) sea surface temperature values for the period 2000–2013 in South East Europe. (**b**) Example of the spatial distribution of the mean monthly (January) chlorophyll *a* values for the period 2003–2013 in South East Europe (Kolios et al. 2015)

More specifically, the user can select the mean monthly distribution of any of the three abovementioned parameters and then the spatial distribution (mean monthly values for a 10- to 12-year period) of the selected parameter for the whole Eastern Mediterranean Sea is provided (Fig. 2.12). Such spatial information is very useful as it highlights sea regions and sea corridors that have high or low values of the abovementioned parameters. These specific areas could be further studied in detail and the corresponding agencies are able to adapt environmentally friendly policies to achieve a sustainable future for the marine and coastal environments in harmonization with the sea transportation and commerce sectors.

2.1.6 Conclusions

This case study presents the main menus and functionalities of the TEN ECOPORT WebGIS platform, a Web-based GIS application that aims to collect and present important environmental information from a network of 12 sea ports in South East Europe. This platform operates as an environmental guide highlighting "hot spots" in port territories and their greater sea areas. It could be used to provide information about the adoption of new practices and policies for a sustainable and environmentally friendly evolution of these ports.

Satellite remote sensing data, as well as ground measurements from stations in ports, are used and plotted through this WebGIS platform and important environmental information regarding the specific 12 port areas is also presented in the WebGIS platform (through the "PORTS" menu). More specifically, information about regions, as well as locations inside ports, of specific environmental interest is easily illustrated through the "Port regions" and "Eco-mapping" submenus, respectively. The differently colored polygons and the "Eco-mapping" symbols highlight official information provided by the involved environmental departments of the ports regarding locations and areas that affect (or are affected by) environmental conditions in port territories. Furthermore, graphs of the temporal evolution of marine/air pollution parameters, as well as the mean spatial distributions of selected marine parameters in small sea areas around ports, are provided (through the "Ports Mapping Parameters" submenu).

Moreover, the TEN ECOPORT WebGIS platform provides (through the "Mapping Parameters" menu) useful and high-quality information about sea corridors for the whole of South East Europe (including also the sea areas around the 12 ports under examination). It visualizes the mean monthly spatial distributions of three important marine parameters (SST, chlorophyll *a*, and CDOM). The provision of such information can be used as an initial guide to study and adopt new legislation and environmentally friendly policies to protect the marine environment, simultaneously taking into consideration the growth of the sea transportation and sea commerce sectors.

From a technical point of view, modern tools and methodologies have been implemented to develop the TEN ECOPORT WebGIS platform following the criteria of portability, flexibility, user friendliness and capacity for updating.

2.2 WebGIS Application for Port Weather Monitoring/ Forecasting

2.2.1 Introduction

The meteorological parameters (such as precipitation, wind speed and wind direction, temperature, humidity, cloudiness, pressure, solar and terrestrial radiation, etc.) are considered some of the most significant physical parameters, because they are essential for the study of a wide variety of phenomena in the atmosphere, the climate, the weather, and the physical environment at the Earth's surface.

The climate change worldwide, the intensification of the greenhouse effect, local disturbances in climatic profiles, severe weather predictions, and extreme weather phenomena (e.g., heat and cold waves, strong winds, cloud storms, heavy precipitation) constitute only a small number of issues where monitoring and forecasting of meteorological parameters is necessary. Such parameters can consequently offer a valuable informational background for sustainable environmental management, future planning, and decision making for the protection of human lives and properties.

With regard to the European transportation sector (assuming all transportation means), it has been reported that, based on the findings of a cost analysis published in 2012 (Nokkala et al. 2012), extreme weather events currently cost the European transport system approximately 15 billion euros annually. The future projections (for the period 2040–2070) show that these large losses will continue to exist, although they will decrease. In terms of maritime transportation, and especially regarding the short sea shipping on which we are focused, the current economic costs are larger than 10 million euros on an annual basis.

In general, extreme weather events are expected to increase in frequency and intensity due to the rapidly changing climate conditions worldwide (IPCC 2014). Such scenarios can affect—among other things—port areas, and many related port activities have to be redesigned for a sustainable and environmentally friendly future of ports and maritime transportation (e.g., Watson 2012; Geels 2012).

Issues regarding transportation as well as many other related procedures are affected by extreme weather events that have great importance for the economic development of sea commerce. Regarding maritime safety and the accuracy of weather information, international organizations, i.e., the International Maritime Organization (IMO) and the World Meteorological Organization (WMO), define legislation and strategies and provide weather reports and forecasts to decrease accidents and economic and human losses.

In conclusion, essential needs in some key aspects regarding weather and sustainable development of sea commerce are summarized as follows:

- Clear and robust prognosis of extreme weather events near ports
- Early warning systems and automatic real-time information exchange systems
- Integration of vessel traffic systems (VTS) and automated identification systems (AIS) with the weather information systems at ports
- Weather forecast quality (especially regarding storms, heavy precipitation, wind, waves, and extreme phenomena)

It is obvious that analytical and continuous recording of meteorological parameters as well as their timely and accurate forecasting are of crucial importance to ports.

2.2.2 The PORTWEATHER WebGIS Application

In the context described above, the PORTWEATHER WebGIS application has been developed (http://www.portweather.eu). It is a useful tool that provides not only marine weather forecasts to support mercantile and passenger traffic but also information concerning past weather events and the climatic profile of four ports: Bari (Italy), Igoumenitsa (Greece), Patras (Greece), and Corfu (Greece).

The PORTWEATHER application was developed in the framework of the pAssengeRs and loGistics information Exchange System (ARGES) project, cofunded by the European Transnational Cooperation Programme Greece–Italy 2007–2013. The ARGES project was based on the cooperation of ten Italian and Greek partners, among which were four port authorities, aiming to enhance the connections between the national information systems and the SafeNet system (http://www.emsa.europa.eu/ssn-main.html), providing—among other things—new services for users, such as marine weather forecasting, a monitoring system for traffic near ports for secure movement of dangerous goods, and a system addressed to passengers including promotional touristic information about the collaborating territories.

2.2.3 Data Used

The PORTWEATHER WebGIS application for marine weather monitoring/forecasting (http://www.portweather.eu) includes real-time measurements from stations near the ports of interest, as well as 3-hourly forecasts for 3 days ahead. Measurements are presented and concerns regarding basic meteorological parameters are forecasted (Table 2.2).

All the real-time and forecasted values for the parameters in Table 2.2 come from the http://openweathermap.org service. Through this service, any user can

Table 2.2 Meteorological parameters used in the PORTWEATHER WebGIS application

Parameter	Unit
Temperature	Degrees Celsius
Precipitation	Millimeters
Wind speed	Beaufort scale
Wind direction	Degrees
Cloudiness	Percentage
Humidity	Percentage

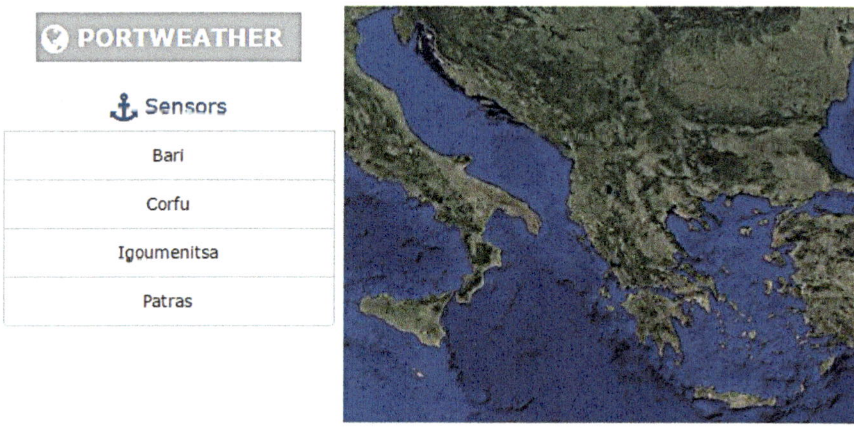

Fig. 2.13 Central interface of the PORTWEATHER WebGIS application (http://www.portweather.eu)

retrieve data from ground weather stations as well as some forecasts. This information is retrieved through APIs at suitable time intervals and in a text format. Next, all the retrieved data are analyzed or/and visualized automatically by the PORTWEATHER WebGIS application, and simultaneously they are stored in the database of the WebGIS application for further future analysis.

2.2.4 Main Menus

The central graphical user interface is provided via the link http://www.portweather.eu. On the left side of the menu, there is a list of presenting ports (Fig. 2.13). When a port of interest is selected, the greater area of the specific port appears.

Using the "Climatic DATA" submenu, the climatic profile regarding the mean monthly temperature and relative humidity of the area around the port is seen through a graphical representation (Fig. 2.14).

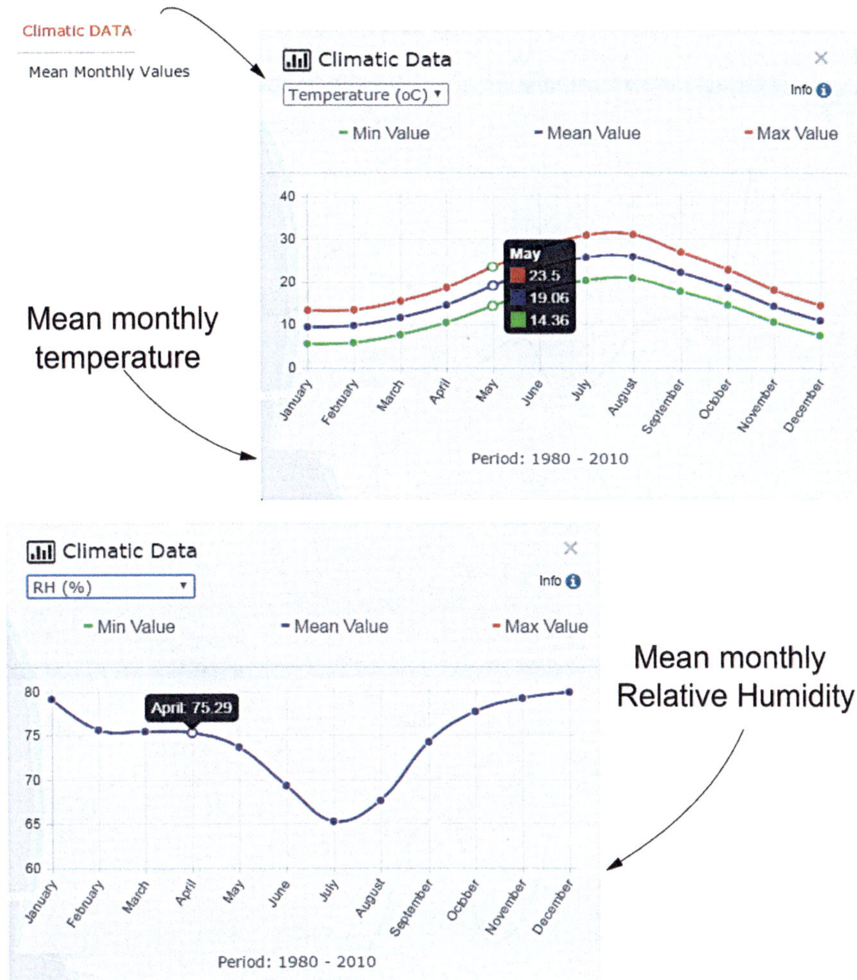

Fig. 2.14 Examples of graphs regarding the climatic monthly variation of basic meteorological parameters

Using the "Historic DATA" submenu, the user can see, through graphs, the temporal evolution of the selected meteorological parameters (i.e., temperature, pressure, sea surface temperature, ground surface temperature, humidity, clouds, wind speed, wind direction, and rain) for a past period of time between two different dates of choice (Fig. 2.15).

The current meteorological situation in the selected port area is visible through the "Current Conditions" submenu. When this menu is selected, a legend containing the current values of several meteorological parameters (temperature, pressure, humidity, clouds, wind) appears (Fig. 2.16).

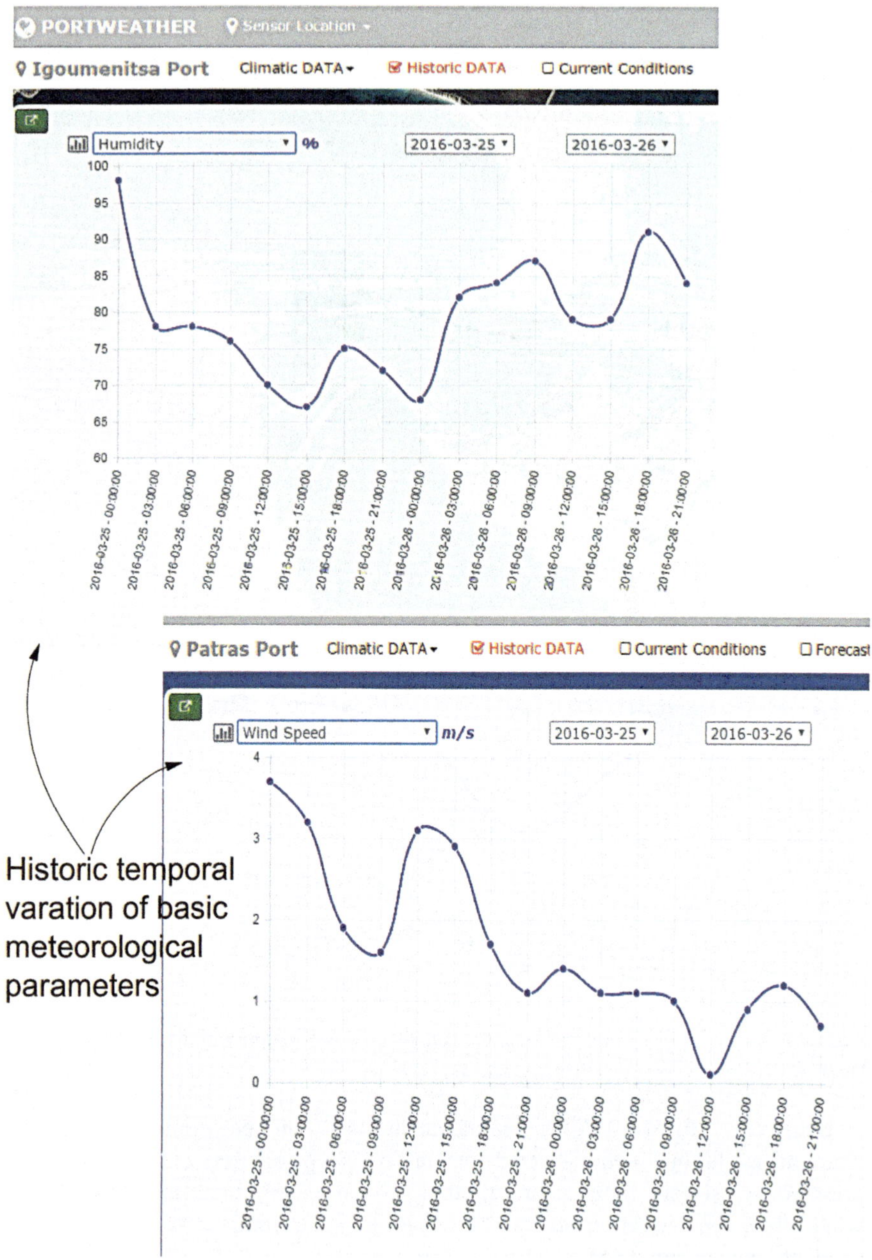

Fig. 2.15 Examples of graphs regarding the historical temporal evolution of basic meteorological parameters

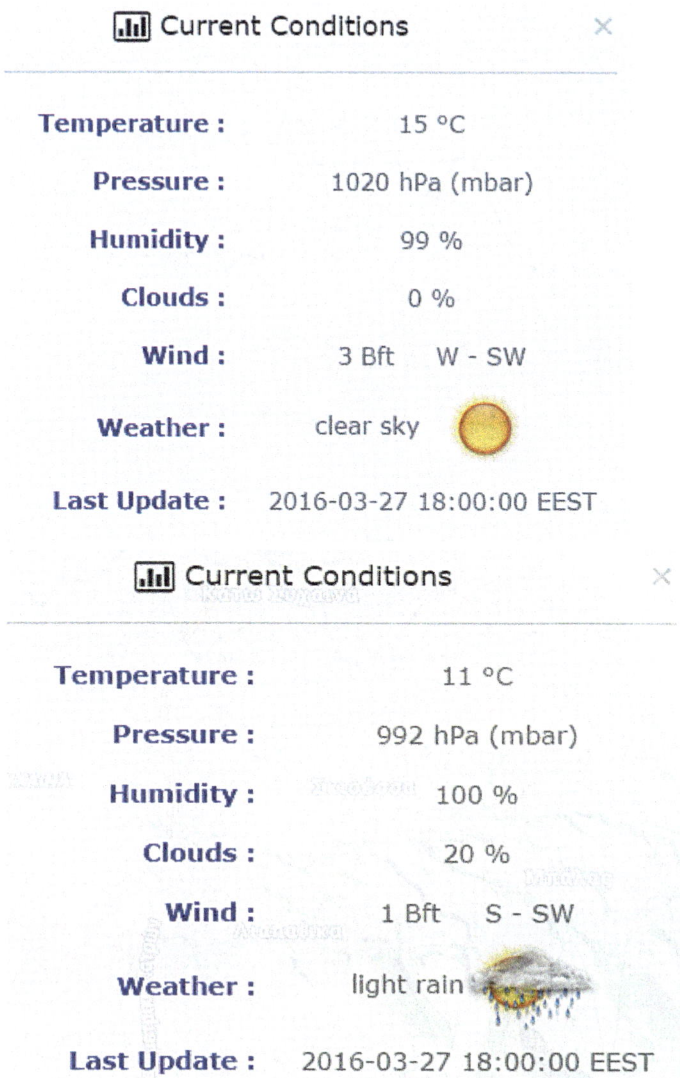

Fig. 2.16 Examples presenting an informative table of current weather conditions in a port area

For all the mentioned parameters (Table 2.2), 3-day forecasts are also provided by selecting the "Forecast" submenu (Fig. 2.17).

In addition, the PORTWEATHER WebGIS application provides interactive maps, which present the spatial distributions of the respective meteorological parameters (Fig. 2.18) when the "Currents," "Waves," "Clouds," "Precipitation," "Wind," and "Temperature" submenus are selected. This module is particularly

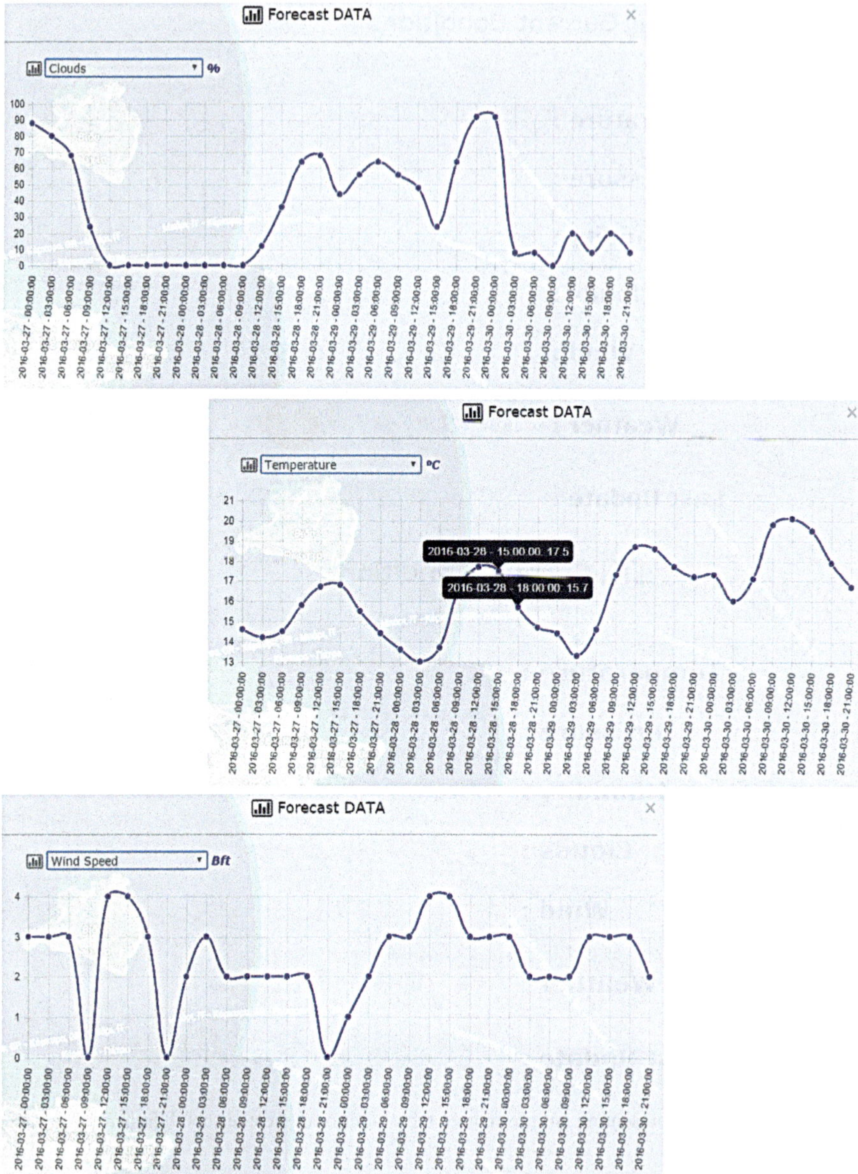

Fig. 2.17 Examples of graphs with 3-hourly forecasts of meteorological conditions in a port area

useful as it provides the weather conditions in the sea corridors that link the main ports in the Adriatic and Ionian Seas.

In addition to this, the PORTWEATHER WebGIS application includes an alert system about extreme weather events. Such weather events are considered of

Fig. 2.18 Spatial distribution of precipitation provided by the http://openweathermap.org service as visualized through the PORTWEATHER WebGIS application

particular interest because they may cause great economic losses, damage to infrastructure, damage to private properties, and even human losses. Thus, alerts about potential extreme weather events are of crucial importance for sea transport and all port activities.

More analytically, when the value of a meteorological parameter (monitored or forecasted) exceeds a threshold value, an alert sign (like the ones presented in Fig. 2.18) appears in the graphical interface of the application. The threshold values for all the meteorological parameters have been defined through a climatic statistical analysis of the historical data for the last decade. There are three different alert labels (signs) according to the intensity of the extreme weather event (Fig. 2.19). The first level of alert refers to a potentially extreme event, the second level is the appearance of an extreme event, and the third level refers to a very extreme event.

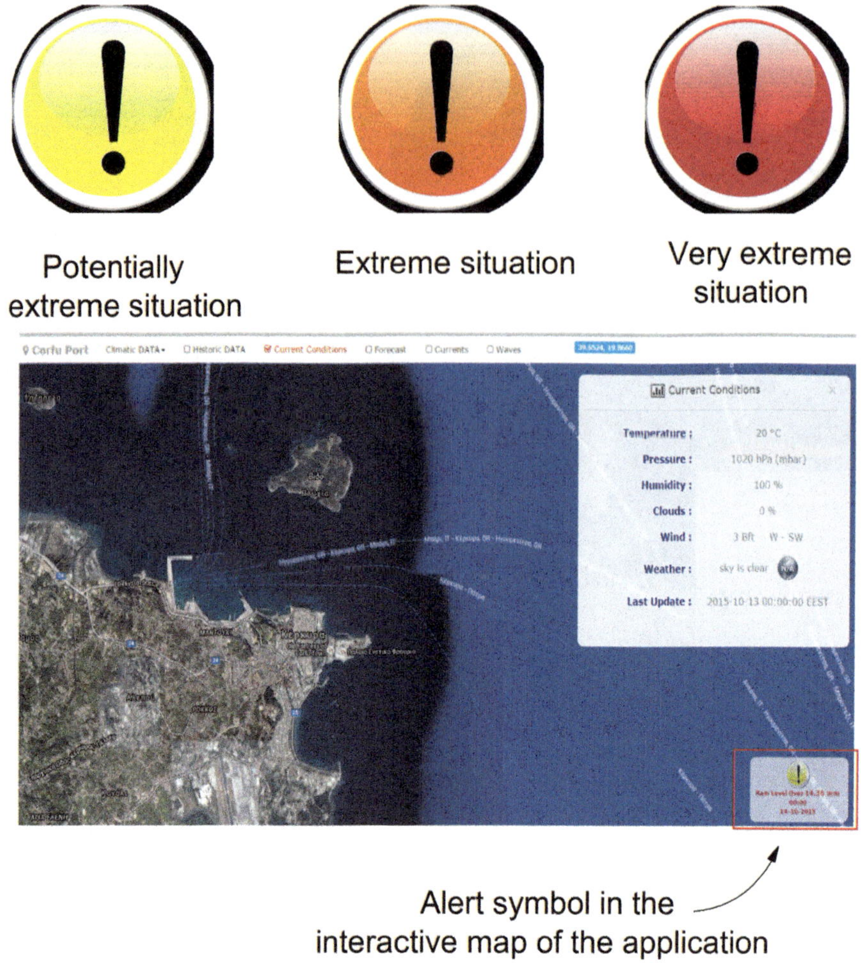

Fig. 2.19 Alert symbols in the PORTWEATHER WebGIS application informing about extreme weather conditions

2.2.5 Conclusions

The presented PORTWEATHER WebGIS application (http://www.portweather.eu) provides essential information about weather conditions in port areas of the Ionian and Adriatic Seas as well as in their greater areas. The application was developed with the use of modern techniques and constitutes an integrated, easy-to-use, portable Web-based interface. This application not only provides information about current meteorological situation and forecasts, but also provides alerts in case of extreme weather events referring to temperature, precipitation, and wind. The PORTWEATHER WebGIS application is very useful for port stakeholders as it helps them to improve the safety of many operations, and especially those activities that are affected by meteorological conditions and extreme weather phenomena.

2.3 WebGIS Application for Extreme Weather Events

2.3.1 Introduction

All extreme weather situations are significantly influenced—more or less—by climatic changes. More specifically, the natural variability of the atmospheric environment, which is important for the formation of extreme weather conditions, is closely related to any climatic change. Nowadays, there is a tendency for more frequent and more intense extreme weather conditions to occur (IPCC 2012).

Studies on extreme weather situations are considered of particular interest because such phenomena can cause considerable economic losses, crop destruction, damage to facilities and private properties, human injuries, and life losses (e.g., Dotzek and Forster 2011; Kober and Tafferner 2009; Kotroni et al. 2011; Papagiannaki et al. 2013).

The effects and kinds of climate change vary depending on the geographic area. Concerning Greece, several studies have already revealed climate tendencies and anomalies, based mainly on statistical methodologies, time series of data, and climate indices and parameters (e.g., Houssos and Bartzokas 2006; Maheras et al. 2006; Nastos and Zerefos 2008; Founda et al. 2004; Kalimeris et al. 2011).

The importance of monitoring, identifying, and predicting extreme weather phenomena is unquestionable. A Web-based geodatabase showing the spatial distribution of such extreme weather events is a valuable tool in order to highlight areas with increased frequency of occurrence. Such information can operate as a proxy to design prevention and protection policies for the population, infrastructure, properties, and mainly human lives in case of extreme weather situation occurrence.

2.3.2 The SevereWeather WebGIS Application

In this section, SevereWeather (http://qgiscloud.com/ko_sta/Weather_extreme_events_Greece) is presented. This is a WebGIS application that provides spatial information about the locations in Greece where extreme weather events (more precisely regarding heavy precipitation, floods, hail, snowfall, and strong winds) have occurred in a 20-year period (1992–2012).

More analytically, this WebGIS application operates as an integrated solution for developing applications of environmental interest using existing GIS software. It is suitable also for users without much skill in programming language. It is based on QGIS software, which is one of the most widely used free and open-source desktop GIS applications worldwide, among a wide range of GIS software choices. It includes many tools to create, edit, visualize, analyze, and publish geospatial information on Windows, Mac, and Linux operating systems. One important characteristic of QGIS is the availability of many plug-ins that continuously expand the data management, analysis, and visualization capabilities.

2.3.3 Collection of Data

The SevereWeather WebGIS application uses data collected from several official reports of organizations, such as the Institute for Environmental Research and Sustainable Development of the National Observatory of Athens (http://www.noa.gr and http://www.meteo.gr) and the Department of Water Resources and Environmental Engineering of the National Technical University of Athens (http://www.itia.ntua.gr), as well as public reports referring to extreme weather events in the Greek territory. In this way, data on a large number of events were collected (578 events in total) referring to a 20-year period (1992–2012). All this information was stored in the geodatabase and GIS-compatible files (vector and raster files) for Web mapping were created.

2.3.4 Stages of Development

The main steps for the development of the SevereWeather WebGIS application were:

1. QGIS Desktop installation (http://www.qgis.org)
2. QGIS Cloud plug-in installation (http://www.sourcepole.com/en/seite-1/qgis-cloud)
3. Sign up in QGIS Cloud (free or by charge)

QGIS Cloud comprises a hosting service for the QGIS Server that is provided as a plug-in in QGIS Desktop. An essential characteristic of QGIS Cloud is the option not only to share data via Web Feature Service (WFS) (see Sect. 1.10) but also to manage them via a transactional Web Feature Service (WFS-T). The basic WFS allows querying and retrieval of features. The WFS-T allows creation, deletion, and updating of features.

The main procedures for the development of the WebGIS application using QGIS Cloud are given in Fig. 2.20.

Describing the procedures shown in Fig. 2.20 more analytically, after the installation of the QGIS software, the user can import and manage data and information of interest. Different types of data can be integrated, labeled, and analyzed so that the final layers of data are produced. It should be mentioned

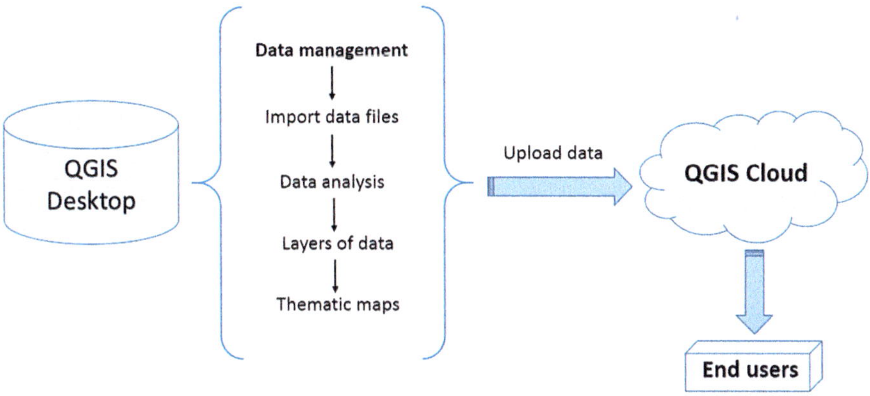

Fig. 2.20 General flow chart for development of a WebGIS application using QGIS Cloud

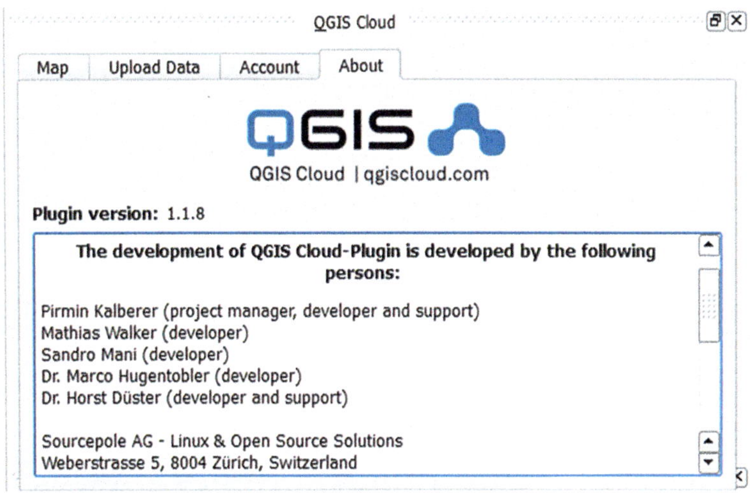

Fig. 2.21 Main menus ("Map," "Upload Data," "Account") of the QGIS Cloud plug-ins in QGIS Desktop (http://qgiscloud.com)

that the data analysis is an optional category of procedures and it depends on the final information chosen to be provided through the WebGIS. Using the data layers of interest, thematic maps can be created in the central interface of QGIS Desktop. These thematic maps usually have a base map of the area of interest and separate layers of data that are overlaid on the base map (e.g., points that represent locations with ground stations, raster files that depict the spatial distribution of a parameter, and so on). Through the relevant menu (Fig. 2.21) of QGIS Cloud, the user must also define a uniform resource locator (URL), which is the address of the Web page for the WebGIS application. The suffix of the URL has the given name of the QGIS Desktop project. The base map or any other data layer can be uploaded separately in

QGIS Cloud using the relevant menus and buttons. Finally, the selected layers are published by using the relevant menu to make them available through the Web interface of the cloud application.

2.3.5 *The Central Interface of the SevereWeather WebGIS Application*

The data layers used for the SevereWeather WebGIS application comprise:

- A detailed interactive base map. The map comes from Bing Maps service, which is a Web mapping provider, part of Microsoft's Bing suite of search engines and powered by the Bing Maps for Enterprise framework.
- A vector file (type: "polygon") with the 13 geographical regions of Greece.
- A vector file (type: "point") with a large number of extreme weather conditions concerning heavy precipitation, floods, hail, snowfall, and strong winds.

These three layers of data were imported into QGIS Desktop (version 2.10.1) (Fig. 2.22).

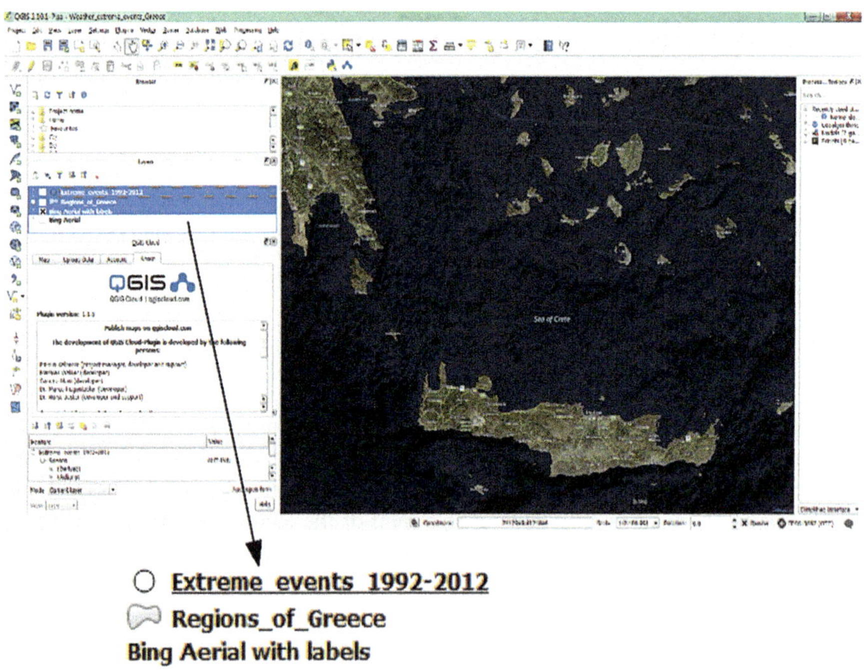

Fig. 2.22 Central interface of QGIS Desktop: the selected information in the *left panel* represents the data layers of interest

When all appropriate analyses and modifications of the selected data layers are fulfilled through QGIS Desktop, the QGIS Cloud plug-in allows publication of the final data layers using the predefined URL.

The central interface of the specific WebGIS application (http://qgiscloud.com/ko_sta/Weather_extreme_events_Greece) with the abovementioned data layers can be seen in Fig. 2.23a.

The central interface provides all the available data layers. In addition, there are tools to zoom in or zoom out in maps and layers, and an object identification tool, as well as tools to measure distance and size (Fig. 2.23b).

More specifically, in the central interface, the data layers "Regions_of_Greece" present the 13 regions of Greece with differently colored polygons (Fig. 2.24). Moreover, the names of the regions are clearly visible.

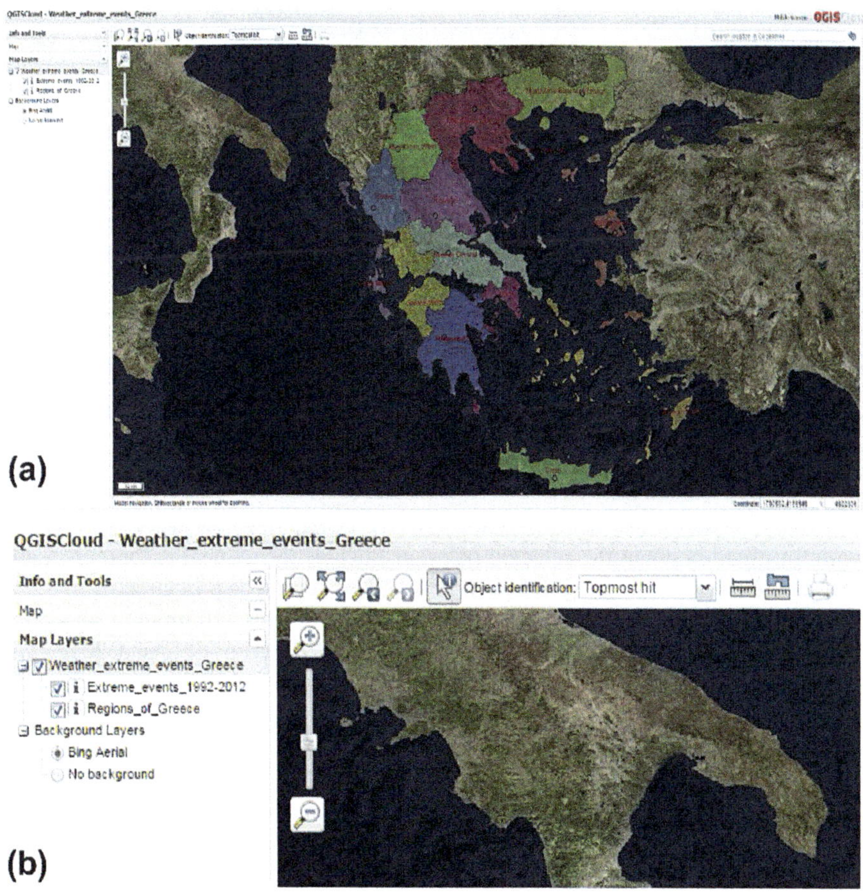

Fig. 2.23 (**a**) Central interface of the WebGIS application for extreme weather events in Greece. (**b**) Snapshot of extreme weather events in the WebGIS central interface: on the *left*, the available data layers; at the *top*, the basic tools provided for this application

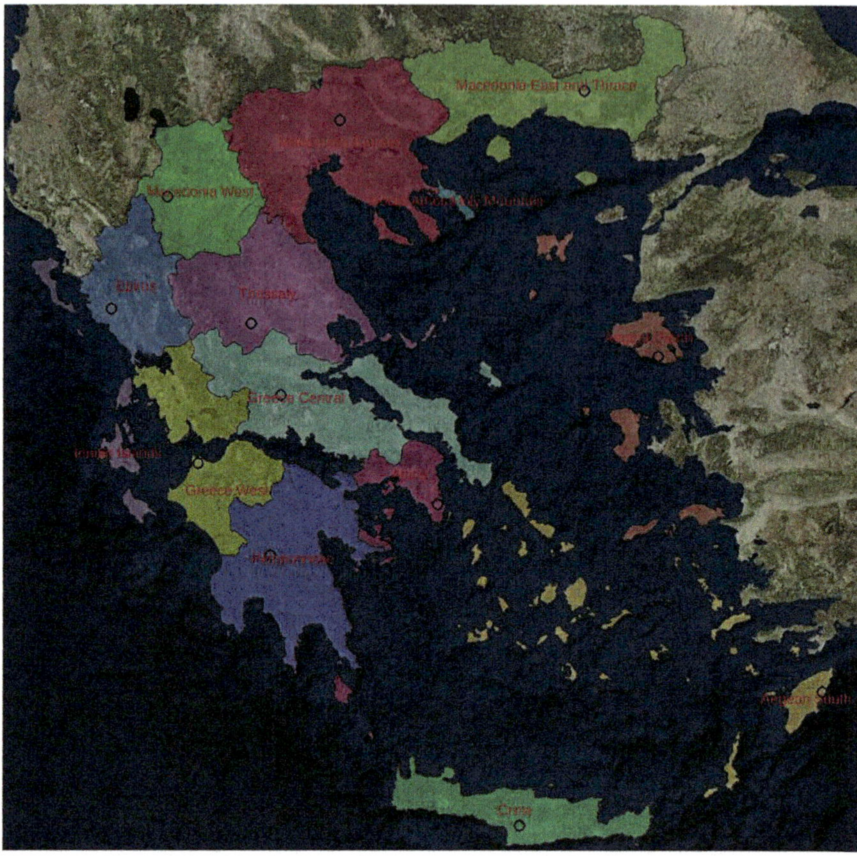

Fig. 2.24 The 13 regions of Greece shown with *differently colored polygons*

When the user clicks on any of these polygons, an attribute table is displayed with information about the selected polygon, such as the name, the total area, and the population according to the latest (year: 2011) census data in Greece.

The data layer "Extreme_events_1992–2012" contains information about the recorded extreme weather events during the period between 1992 and 2012. The extreme events are given per region and per date. The user, by clicking on the black circle inside a region, can see all the recorded extreme weather events in the region as well as relevant information about them (type of event, date of event, year, season, and intensity). Figure 2.25 presents a characteristic example.

It is worth noting that the application provides the choice to operate the mobile version by clicking on the "Mobile Version" link in the top right corner of the central interface. This capability is very important because the graphical user interface is automatically adjusted to mobile devices, thus providing the end user with a fully functional Web-based application accessible from any device.

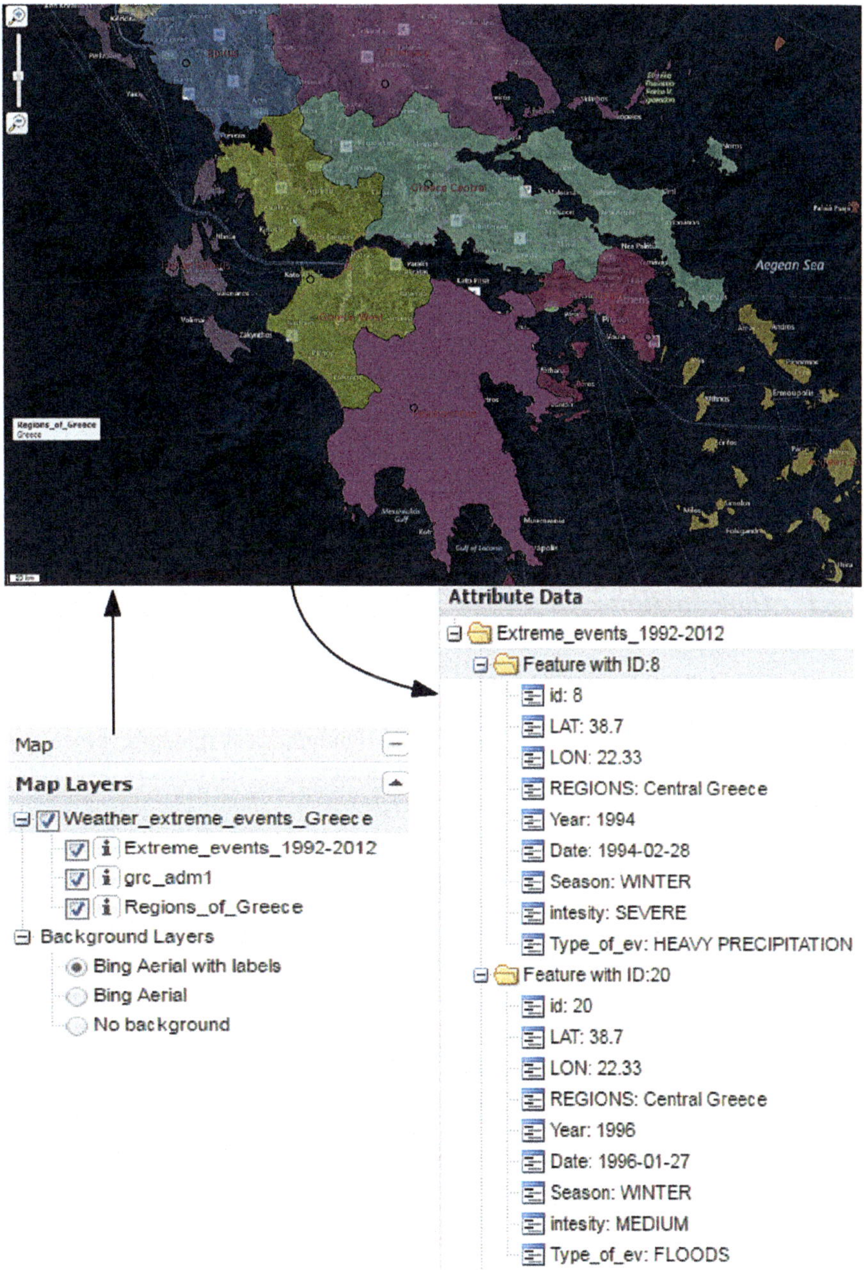

Fig. 2.25 Characteristic example of the attribute table of the application

2.3.6 Conclusions

As was already mentioned in Chap. 1 (Sect. 1.14), the solution of cloud GIS can be an alternative modern way to develop and use Web-based environmental GIS applications allowing information to be easily accessible, thus largely contributing to decision support systems regarding natural protection and early warning for natural hazards. In this section, we present a characteristic example of a cloud GIS application regarding cases of extreme weather event occurrence per geographic region in Greece. It is a simple and modern application, and is easy to develop, use, and update. Generally speaking, using cloud GIS, new perspectives can be created as the information is provided to end users through the Internet and mobile devices, such as smartphones and tables.

2.4 WebGIS Application for Satellite Weather Monitoring/Forecasting

2.4.1 Introduction

Cloud storm development has major and direct impacts on human lives and properties. It is known that cloud storms produce severe weather conditions, such as heavy rain, hail, strong winds, tornadoes, lightning, and flooding, which can significantly impact human activities (e.g., Maddox et al. 1986; Romero et al. 2000; Rutledge et al. 1993; Gaye et al. 2005; Correoso et al. 2006). The forecasting of these events is very difficult and extremely complicated not only due to their small-scale internal dynamics, but also because they are developed under various favorable conditions depending on topography, synoptic weather conditions, the humidity of the atmosphere, atmospheric instability, wind shear, and many other factors.

The use of modern geostationary meteorological satellites with their fine time sampling (typically, 15 min) and space sampling (3 km at the subsatellite point), as well as their large geographical coverage, constitutes an excellent alternative way to face the uncertainness and the restrictions of numerical models based on radar for satellite monitoring/forecasting of cloud storms. There are several approaches worldwide to monitor storms and their evolution, especially in areas with important storm activity like the greater Mediterranean Basin (e.g., Riosalido et al. 1998; Morel and Senesi 2002; Kolios and Feidas 2012a, b).

In the last few years, the use of GIS capabilities combined with modern Web technologies has led to the development of automated systems, which use satellite images as data inputs and provide useful information regarding storm evolution through a WebGIS interface.

A characteristic example of such integrated systems based on satellite datasets to monitor/forecast cloud storms through Web-based applications is the ForTraCC system (Vila et al. 2008), which refers to the tracking and forecasting of cloud storms over South America (http://sigma.cptec.inpe.br/sosmanaus) based on information from the Geostationary Operational Environmental Satellite system (GOES satellites). Another characteristic example is the fully automated system called NEFODINA (Puca et al. 2005), which focuses over the Italian peninsula and its surroundings, based on multispectral Meteosat Second Generation (MSG) imagery (http://www.meteoam.it/nefodina/en).

2.4.2 The SatWeather WebGIS Application

In the following section, an experimental WebGIS application for cloud storm monitoring/forecasting is presented. The application is called SatWeather (http://satweather.kic.teiep.gr) and has been developed to visualize cloud storm patterns as well as major physical phenomena caused by their existence (precipitation, hail, lightning), using modern Web technology tools and features. At this point, it must be mentioned that all the principles, methods, and techniques of this application are exclusively based on the PhD thesis of Kolios (2009). This application practically visualizes, in a modern Web-based graphical user interface, all the final products extracted from the latest version of the system developed by Kolios (2009), where—among others—hail and lightning estimation as well as changes in the cloud tracking procedure, are included. More specifically, the application is focused on monitoring the cloud tops of storms (and/or cloudy areas that can evolve into storms in the following few hours) in the greater Mediterranean Basin, using exclusively satellite images. A schematic chart of the data flow of the application can be seen in Fig. 2.26.

The satellite images are provided directly by the Meteosat satellite (http://www.eumetsat.int). The initial multispectral imagery (Table 2.3) is automatically preprocessed and the images are analyzed on a pixel basis in the main system, following all the analytic methods and procedures of Kolios (2009) as well as Kolios and Feidas (2012a, b) to detect and monitor cloud tops of storms. All these procedures are used to operationally provide monitoring of the cloud top evolution as well as forecasts up to 6 h ahead (short-range forecasting).

As mentioned before, four types of information are provided through this Web-based application on a pixel basis, either in monitoring or in the forecasting module:

- Cloud top temperature (°C)
- Precipitation rate (mm/h)
- Hail probability (%)
- Lightning probability (%)

The typical temporal resolution of the provided information is 15 min, as it is the data flow of the Meteosat satellite.

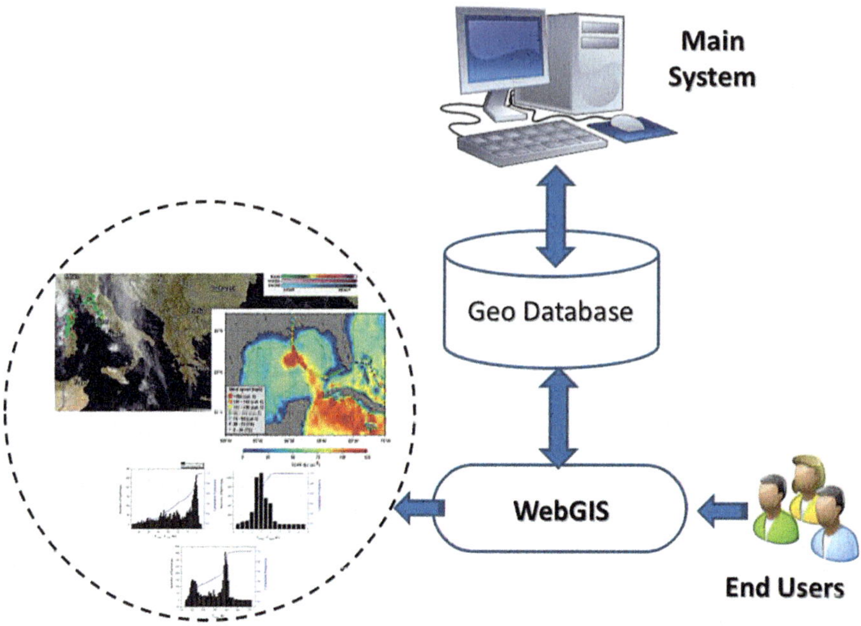

Fig. 2.26 The final products from the latest version of the core algorithm developed by Kolios (2009), are used from the "Main system" of the SatWeather application. All the products which automatically collected from this core algorithm, organized in a central database (Geo Database). All the collected information is finally visualized using the map tiles approach and provided to the end users through the interactive maps of the SatWeather WebGIS interface (http://satweather.kic.teiep.gr)

Table 2.3 Basic characteristics of Meteosat multispectral imagery used by the application

Channel	Spectral width (μm)	Spectral center (μm)
5	5.35–7.15	6.2
6	6.85–7.85	7.3
7	8.3–9.1	8.7
9	9.8–11.8	10.8
10	11–13	12.0

2.4.3 Characteristics and Visualization of the SatWeather Application

The central interface of the SatWeather WebGIS application shows an interactive map of the Earth globe (Google map) but it focuses on and provides information about the greater Mediterranean Basin (Fig. 2.27).

Figure 2.27 presents the central interface with the interactive map. In the top right corner of the interface, the user can choose one or more of the parameters available for monitoring/forecasting, i.e., clouds, rain, hail, lightning. Using this

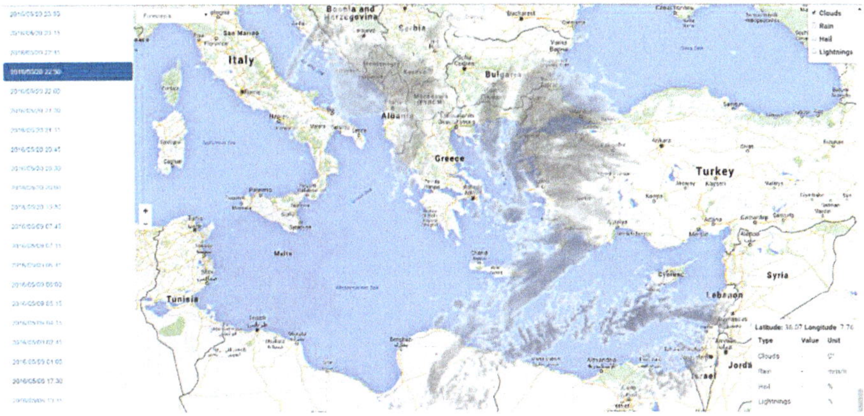

Fig. 2.27 Central interface of the SatWeather WebGIS application

menu, the interactive map of the greater Mediterranean Basin presents the results of the selected parameter(s). When the small menu/checklist box named "Forecasts" is chosen in the top left corner of the map, a series of dates and times ahead, i.e., the available forecasts, can be seen. When one of the available date/time options is chosen, a map of the forecasted parameters of interest can be seen. When the cursor is moved on the map, the bottom right stable table shows the values of the parameters pixel by pixel corresponding to the specific coordinates of the cursor. Nevertheless, if the user wants to see all the available information for a specific pixel and it is difficult to locate it just by moving the cursor, the zoom button can be used, and then when a specific point of interest on the enlarged map is clicked on, all the available information appears. Finally, on the left side of the Google map there is a list of past dates/times with the available information. When one of the past dates/times is chosen, it is considered the current date/time and all the available forecasts for this date/time can be seen using the "Forecasts" menu.

As mentioned, the SatWeather WebGIS application also includes capabilities for hail/lighting estimation. Figures 2.28 and 2.29 present illustrative examples of how this information can be seen through the WebGIS interface. The user just has to select the appropriate checkboxes on the right of the interface and the relevant information is presented for the time/date, which is active on the left side of the interface (or the forecast for the time/date of interest).

2.4.4 Conclusions

The SatWeather WebGIS application for monitoring and forecasting cloud storms (http://satweather.kic.teiep.gr) has been designed and is operating on a near real-time basis. It uses the map tiles solution for Web mapping and constitutes an automated Web interface where basic meteorological data products are visualized

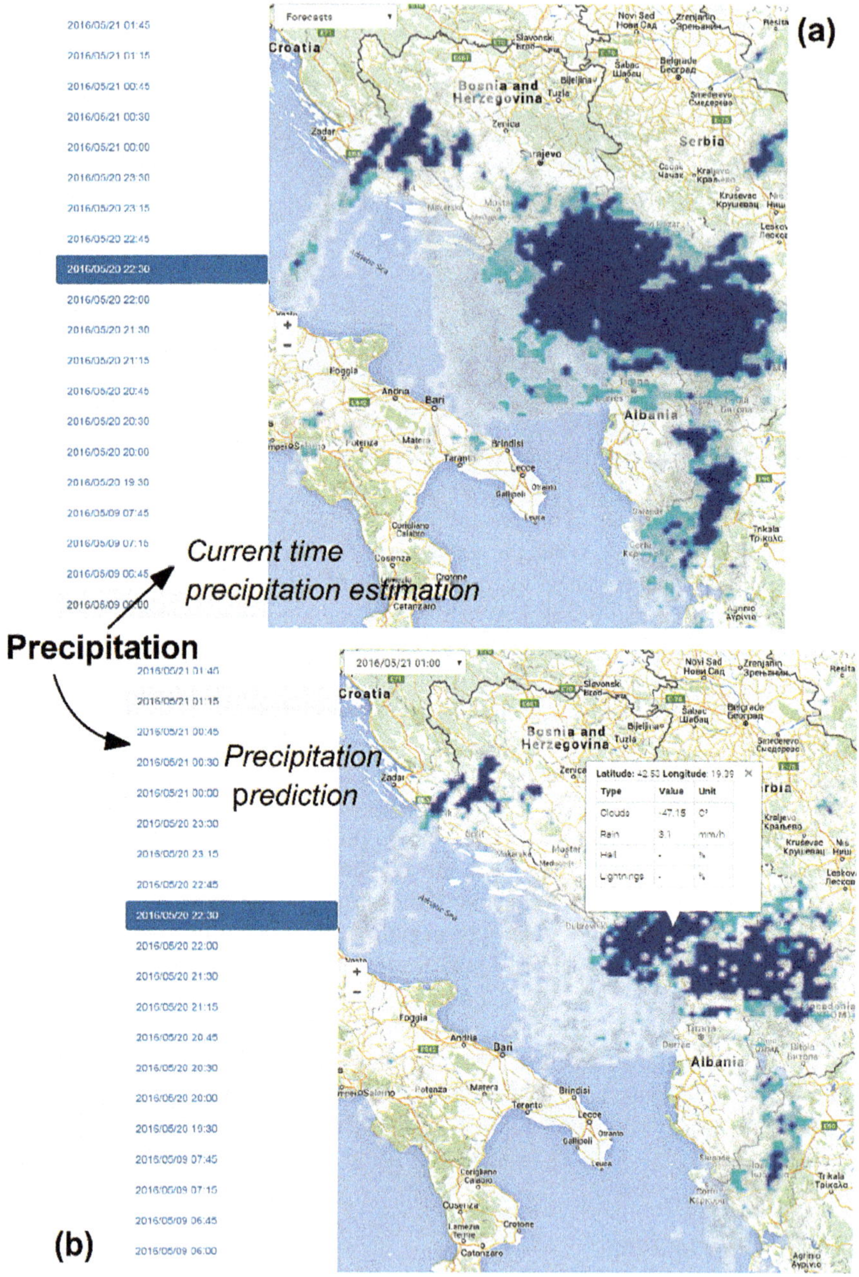

Fig. 2.28 (**a**) Cloud storms at 22:30 h (UTC) on May 20, 2006, in the province of Epirus (Greece), northern Albania, and Montenegro. The *blue-colored cloud areas* indicate areas with estimated precipitation. (**b**) Cloud storms predicted to occur at 01:00 h (UTC) on May 21, 2016, based on the available information 135 min before (at 22:30 h (UTC) on May 20, 2016). The *blue-colored cloud areas* indicate the areas where it was predicted to rain

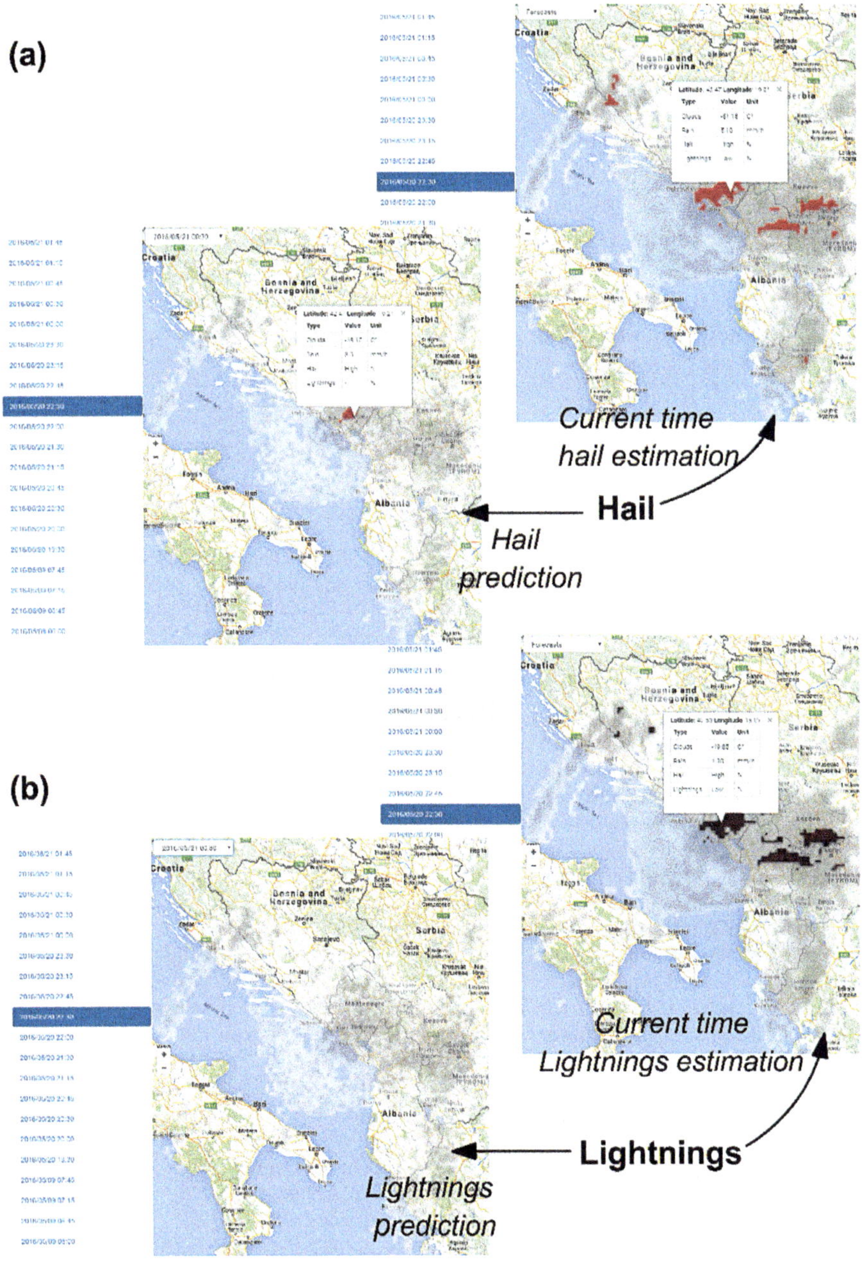

Fig. 2.29 (**a**) Example of the hail estimation (*red coloured areas*) at 22:30 h (UTC) on 20th May, 2016 as well as the hail forecast (hail predicted to occur) 00:30 h (UTC) on 21st May, 2016 using information 120 min before (22:30 h UTC on 20th May, 2016) (**b**) The same as the hail but for the lightning events

in time and space. These data come from the Meteosat satellite multispectral imagery. After preprocessing and analyses, the data are transformed into useful information for monitoring and short-range forecasting of cloud storms and their potential results (e.g., precipitation, rain, lightning). The developed application combines Web technologies, programming, and satellite remote sensing techniques in order to provide, in a simple way, timely, accurate, and useful information for public agencies, the agriculture sector, companies, and any other organizations whose activities and responsibilities can be affected by extreme weather conditions.

At this point, it is worth pointing out that evaluation of the performance is provided mainly for accuracy of the forecasting/monitoring procedures and products. Also, new parameters and procedures could be added.

References

Bailey D, Solomon G (2004) Pollution prevention at ports: clearing the air. Environ Impact Assess Rev 24(7–8):749–774

Becker A, Inoue S, Fischer M, Schwegler B (2012) Climate change impacts on international seaports: knowledge, perceptions, and planning efforts among port administrators. Clim Change 110(1):5–29

Brewin R, Raitsos D, Pradhan Y, Hoteit I (2013) Comparison of chlorophyll in the Red Sea derived from MODIS-Aqua and in vivo fluorescence. Remote Sens Environ 136:218–224

Correoso FJ, Hernandez E, Garcia-Herrera R, Barriopedro D, Paredes D (2006) A 3-year of cloud-to-ground lightning flash characteristics of mesoscale convective systems over the Western Mediterranean Sea. Atmos Res 79:89–107

Dinwoodie J, Tuck S, Knowles H, Benhin J, Sansom M (2012) Sustainable development of maritime operations in ports. Bus Strateg Environ 21(2):111–126

Donnelly A, Jones M, O'Mahony T, Byrne G (2007) Selecting environmental indicator for use in strategic environmental assessment. Environ Impact Assess Rev 27(2):161–175

Dotzek N, Forster C (2011) Quantitative comparison of METEOSAT thunderstorm detection and nowcasting with in situ reports in the European Severe Weather Database (ESWD). Atmos Res 100:511–522

Edoho FM (2008) Oil transnational corporations: corporate social responsibility and environmental sustainability. Corp Soc Respon Environ Manag 15:210–222

Eyring V, Isaksen ISA, Berntsen T, Collins WJ, Corbett JJ, Endresen O, Grainger RG, Moldanova J, Schlager H, Stevenson DS (2010) Transport impacts on atmosphere and climate: shipping. Atmos Environ 44(37):4735–4771

Ferreira JG, Vale C, Soares CV, Salas F, Stacey PE, Bricker SB, Silva MC, Marques JC (2007) Monitoring of coastal and transitional waters under EU Water Framework Directive. Environ Monit Assess 135:195–196

Founda D, Papadopoulos KH, Petrakis M, Giannakopoulos C, Good P (2004) Analysis of mean, maximum and minimum temperature in Athens from 1897 to 2001 with emphasis on the last decade: warm events and cold events. Glob Planet Chang 14(1–4):27–38

Fustes D, Cantorna D, Dafonte C, Arcay B, Iglesias A (2014) A cloud-integrated web platform for marine monitoring using GIS and remote sensing. Application to oil spill detection through SAR images. Futur Gener Comput Syst 34:155–160

Gaye A, Viltard A, De Felice P (2005) Squall lines and rainfall over Western Africa during 1986 and 1987. Meteorog Atmos Phys 90:215–224

Geels FW (2012) A socio-technical analysis of low-carbon transitions: introducing the multi-level perspective into transport studies. J Transp Geogr 24:471–482

Grifoll M, Jordà G, Espino M, Romo J, García-Sotillo M (2011) A management system for accidental water pollution risk in a harbour: the Barcelona case study. J Mar Syst 88(1):60–73
Gupta AK, Gupta SK, Patil R (2005) Environmental management plan for port and harbour projects. Clean Techn Environ Policy 7(2):133–141
Houssos EE, Bartzokas A (2006) Extreme precipitation events in NW Greece. Adv Geosci 7:91–96
Intergovernmental Panel on Climate Change (IPCC) (2014) Impacts, adaption and vulnerability. In: Fifth assessment report
IPCC (2012) Managing the risks of extreme events and disasters to advance climate change adaptation. A Special Report of Working Groups I and II of the Intergovernmental Panel on Climate Change (IPCC), Cambridge University Press
Kalimeris A, Founda D, Giannakopoulos C, Pierros F (2011) Long term precipitation variability in the Ionian Islands (Central Mediterranean): climatic signal analysis and future projections. Theor Appl Climatol 109:51–72
Karydis M, Kitsiou D (2012) Eutrophication and environmental policy in the Mediterranean Sea: review. Environ Monit Assess 184:4931–4984
Kober K, Tafferner A (2009) Tracking and nowcasting of convective cells using remote sensing data from radar and satellite. Meteorol Z 1(18):075–084
Kolios S (2009) Development of automated system for the very short range forecast of mesoscale convective systems using Meteosat satellite imagery. PhD thesis, http://hdl.handle.net/10442/hedi/23860
Kolios S, Feidas H (2012a) An automated nowcasting system of mesoscale convective systems for the Mediterranean basin using Meteosat imagery. Part I: System description. Meteorol Appl 20:287–295. doi:10.1002/met.1282
Kolios S, Feidas H (2012b) An automated nowcasting system of mesoscale convective systems for the Mediterranean basin using Meteosat imagery. Part II: Verification statistics. Meteorol Appl 20:296–307. doi:10.1002/met.1281
Kolios S, Stylios C (2015) Coastal marine environment monitoring using satellite data derived from MODIS instrument. Sustainable development of sea-corridors and coastal waters. Special Edition (conference proceedings), Springer, doi:10.1007/978-3-319-11385-2_14
Kolios S, Stylios C, Petunin A (2015) A WebGIS platform to monitor environmental conditions in ports and their surroundings in South Eastern Europe. Environ Monit Assess 187:574. doi:10.1007/s10661-015-4786-x
Kotroni V, Lagouvardos K, Retalis A (2011) The heat wave of June 2007 in Athens, Greece—Part 2: Modeling study and sensitivity experiments. Atmos Res 100(2011):1–11
Kulkarni AT, Mohanty J, Eldho TI, Rao EP, Mohan BK (2014) A web GIS based integrated flood assessment modeling tool for coastal urban watersheds. Comput Geosci 64:7–14
Li A, Bo Y, Zhu Y, Guo P, Bi J, He Y (2013) Blending multi-resolution satellite sea surface temperature (SST) products using Bayesian maximum entropy method. Remote Sens Environ 135:52–63
Maddox RA, Howard KW, Bartles DL, Rodgers DM (1986) Mesoscale convective complexes in middle latitudes. In: Ray PS (ed) Mesoscale meteorology and forecasting. American Meteorological Society, Boston
Maheras P, Flokas H, Tolika K, Anagnostopoulou C, Vafiadis M (2006) Circulation types and extreme temperature changes in Greece. Clim Res 30:161–174
Marinski J, Marinov D, Branca T, Mali M, Floqi T, Stylios C, Damiani L (2015) Guidelines for elaboration management action plan for ecologically sustainable development and management of SEE seaports of trans-European transport networks. Sustainable development of sea-corridors and coastal waters, special edition (conference proceedings), Springer, pp 135–141
Morel C, Senesi S (2002) A climatology of mesoscale convective systems over Europe using satellite infrared imagery. I: Methodology. Q J R Meteorol Soc 128:1953–1971

Nastos PT, Zerefos CS (2008) Decadal changes in extreme daily precipitation in Greece. Adv Geosci 16:56–62
Ng AKY, Liu JJ (2014) Port-focal logistics and global supply chains. Palgrave Macmillan, New York
Nokkala M, Leviakangkas P, Oiva K (2012) The cost of extreme weather events for the European transport system. EWENT Project D4. Available at: http://www.vtt.fi/inf/pdf/technology/2012/T36.pdf
Noyon V, Devogele T (2005) Ships monitoring with distributed GIS for better safety policy. IEEEXplore, Oceans 2005—Europe, vol 1, pp 231–232. doi:10.1109/OCEANSE.2005.1511716
Nursey–Bray M, Blackwell B, Brooks B, Campbell M, Goldsworthy L, Pateman H, Rodrigues I, Roome M, Wright JT, Francis J, Hewitt CL (2013) Vulnerabilities and adaptation of ports to climate change. J Environ Plan Manag 56(7):1021–1045
Papagiannaki K, Lagouvardos K, Kotroni V (2013) A database of high-impact weather events in Greece: a descriptive impact analysis for the period 2001–2011. Nat Haz Earth Syst Sci 13:727–736
PPRISM (2012) Port Performance Indicators: Selection and Measurement (PPRISM). Project Executive Report. Available at: http://pprism.espo.be/LinkClick.aspx?fileticket=-sv7LYCzIJs%3D&tabid=3555
Portman ME (2014) Visualization for planning and management of oceans and coasts. Ocean Coast Manag 98:176–185
Puca S, Biron D, De Leonimbus L, Melfi D, Rosci P, Zauli F (2005) A neural network algorithm for the nowcasting of severe convective systems. In: CIMSA 2005—IEEE international conference on computing intelligence for measurement system applications, Giardini Naxos, 20–22 July, Italy
Puig M, Wooldridge C, Darbra RS (2014) Identification and selection of environmental performance indicators for sustainable port development. Mar Pollut Bull 81(1):124–130
Quynh LX, Hens L, Stoyanov S (2011) Water management in the framework of environmental management systems in Bulgarian seaports. Phys Chem Earth 36:141–149
Riosalido R, Carretero O, Elizaga F, Martin F (1998) An experimental tool for mesoscale convective systems nowcasting. In: SAF training workshop on nowcasting and very short range forecasting, Madrid, Spain, 9–11 December, pp 127–135
Romero R, Doswell CA, Ramis C (2000) Mesoscale numerical study of two cases of long-lived quasi-stationary convective systems over eastern Spain. Mon Weather Rev 128:3731–3751
Rutledge SA, Williams RE, Petersen AW (1993) Lightning and electrical structure of mesoscale convective systems. Atmos Res 29:27–53
Shang S, Lee Z, Wei G (2011) Characterization of MODIS-derived euphotic zone depth: results for the China Sea. Remote Sens Environ 115:180–186
Urquhart E, Hoffman M, Murphy R, Zaitchick B (2013) Geospatial interpolation of MODIS-derived salinity and temperature in the Chesapeake Bay. Remote Sens Environ 135:167–177
Vila AD, Machado AL, Laurent H, Velasco I (2008) Forecast and tracking the evolution of cloud clusters (FORTRACC) using satellite infrared imagery: methodology and verification. Weather Forecast 23:233–245
Watson M (2012) How theories of practice can inform transition to a decarbonized transport system. J Transp Geogr 24:488–496
WHO World Health Organization (2000) Air quality guidelines for Europe, 2nd edn. In: WHO Regional Publications, European Series No. 91 WHO, Copenhagen

Part II
Theoretical Background of the Geomagnetic Field and GIS Applications

Chapter 3
Mathematical Methods of Geomagnetic Field Modeling

3.1 Introduction to the Geomagnetic Field and Its Role

The geomagnetic field, as the Earth's magnetic field is called, extends from Earth's interior to outer space. The magnetosphere is the region defined by the areal extent and the shape of the magnetic field of a celestial body, like the planet Earth. The Earth's magnetosphere extends several tens of thousands of kilometers into space beyond the ionosphere. The magnetosphere protects the Earth from the charged particles emitted by the Sun (the so−called *solar wind*) and the cosmic rays that would, otherwise, strip away the upper atmosphere, including the ozone layer that protects the Earth from harmful ultraviolet radiation.

The spatiotemporal variations of the geomagnetic field extend over a wide range of time and spatial scales depending on the location, the season, the time period, and the interaction with solar wind. Characteristically, they refer to the regular daily variation in the Earth's magnetic field, which also exhibits irregular disturbances, and when these are large, they are called *magnetic storms*. These disturbances are caused by interaction of the solar wind, and disturbances therein, with the Earth's magnetic field. Although irregular, magnetic disturbances exhibit some patterns in their frequency of occurrence. The main pattern is the correlation with the 11−year solar cycle. Also, the amplitude of magnetic disturbances is larger at higher latitudes.

It is worth pointing out the key role of the geomagnetic field in influencing flora and fauna, and in general its vital importance for life on Earth. The geomagnetic field not only protects us from hazardous parts of the solar radiation spectra, but it has also been reported that Earth's magnetic field causes intensification of protein synthesis and disintegration in plant roots (e.g., Belyavskaya 2004; Galland and Pazur 2005). Regarding fauna, many mammals, such as whales and dolphins, as well as many birds, use the Earth's magnetic field to navigate during their seasonal migrations (e.g., Dubrov 1978; Wiltschko and Wiltschko 2005).

S. Kolios et al., *GIS and Environmental Monitoring*, Geotechnologies and the Environment 20, DOI 10.1007/978-3-319-53086-4_3

The vital role of the geomagnetic field for the physical environment on Earth constitutes an important research field with strong potential. It has to interact with and integrate a variety of different sciences so as to cover the needs and requirements for a modern analytic study of the role of the geomagnetic field in Earth's life. With regard to this aspect, the spatiotemporal variations of the geomagnetic field can nowadays be efficiently studied using powerful computer clusters, new models, and integrated systems, such as geographical information systems (GIS), where these variations can be visualized and analyzed and/or combined with many other types of information in a very realistic and representative way.

3.2 Origin of Earth's Magnetism

According to new findings, the Earth's outer core is liquid and mainly metallic. Its main chemical elements are iron and nickel, being at a depth of more than 670 km. These elements respectively make up ~85.5% and ~5.2% of Earth's core mass fraction, i.e., a total of more than 90% (Pushcharovskii 2001; Pushcharovskii and Pushcharovskii 1999; Gavrilov 2005; Belousov 1976; Allegre et al. 1995; McDonough 1999; Golicyn 1991).

Permanent rotations of both the Earth and its core cause constant flows inside its core and consequent electric currents. They, in turn, provide the existence of the geomagnetic field according to the principles of magnetic hydrodynamics (Golicyn 1991; McDonough and Sun 1995; Merril et al. 1996; Weinstein 1983).

If it was simply assumed that the Earth is alone in space, then the planet's magnetic field lines would be represented as symmetrical arcs extending from the north magnetic pole to the south magnetic pole (Fig. 3.1). In this model, the value of magnetic intensity inversely correlates with the distance from the landscape to the observation location.

But, actually, the Earth's magnetic field permanently interacts with the magnetic fields that are generated by the Sun, the planets of the solar system, and other celestial bodies and systems. As a result of this interaction, there is a significant deformation of the geomagnetic field itself. This fact results in some cardinal corrections to the laws describing the Earth's magnetic field ideal model.

It has to be mentioned that this deformed geomagnetic field constitutes a kind of protective shield, which prevents any solar wind penetration into the geospace. Solar wind is the flow of ionized particles (mostly helium–hydrogen plasma), which radially flows from the solar corona to outer space. The solar wind interacts with the geomagnetic field barriers, flows around the Earth, and creates a special region around the geomagnetic field. In the literature (Pudovkin et al. 1975) this region is known as the Earth's magnetosphere, and it is illustrated in Fig. 3.2.

The typical teardrop shape of the magnetosphere is explained by the balance of the solar wind dynamic pressure and coronal plasma flow heat on the one hand, and the Earth's magnetic field pressure on the other hand.

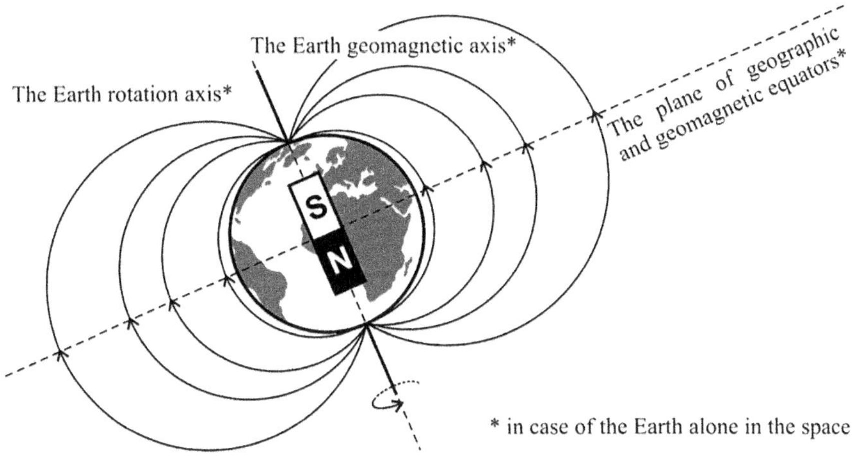

Fig. 3.1 Idealized presentation of the Earth's magnetic field (Reproduced from Vorobev 2013a)

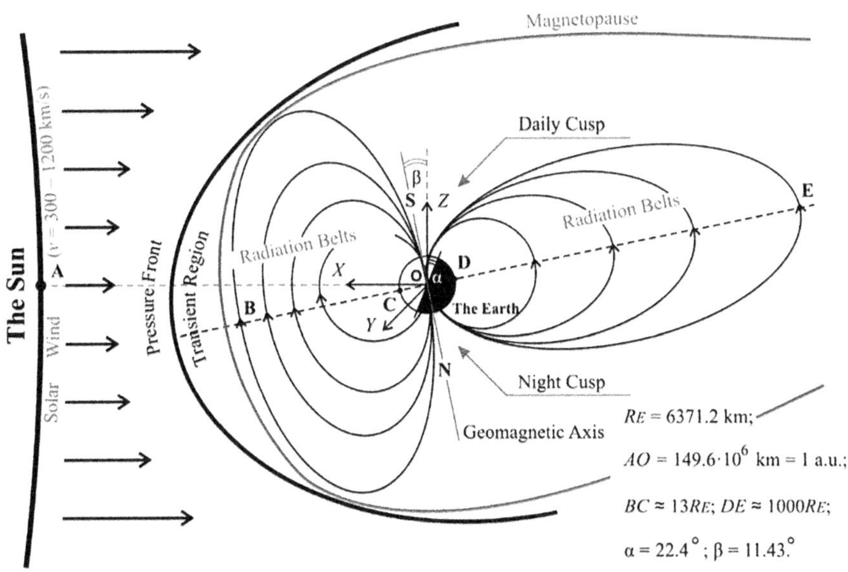

Fig. 3.2 Overview of the Earth's magnetosphere structure

It becomes obvious that, due to the complicated and heterogeneous structure of the geomagnetosphere, its field characteristics are distributed over the landscape and in outer space in an extremely unequal way.

3.3 Geomagnetic Variations and Their Classification

Geomagnetic variations are the declinations of the observed amplitude parameters of the Earth's magnetosphere from the calculated/estimated values that are defined in the normal (undisturbed) geomagnetosphere state.

The following analytical review of the geomagnetic variations is based on the results of many years of observations of the Earth's magnetosphere. The information provided here has been published by the following scientific institutes, organizations, and laboratories:

- Institute of Terrestrial Magnetism, Ionosphere and Radio Wave Propagation of the Russian Academy of Sciences "IZMIRAN", Troitsk, Russia
- GEOMAGNET, Russia
- Tomsk National Research State University, Tomsk, Russia
- National Oceanic and Atmospheric Administration (NOAA), Rockville, USA
- National Aeronautics and Space Administration (NASA), Washington, DC, USA
- British Geological Survey (BGS), Edinburgh, UK

The geomagnetic field is not static and its size is continuously changing. The most essential events and processes (of natural origin), which are usually followed by geomagnetic variations, are described below.

Diurnal Variations of the Geomagnetic Field The Earth's rotation around its own axis causes *diurnal variations of the geomagnetic field* on its surface. These variations are connected with the change of position point from the stationary solar–magnetospheric coordinate system, where the OX axis is directed from the Earth to the Sun; the OY axis is placed in the ecliptic area and is directed from the morning to the evening; and the OZ axis is directed to the ecliptic pole (Fig. 3.2) (Egeland et al. 1976).

Moon–Related Geomagnetic Variations The rotation of the Moon around the Earth causes weak and almost insignificant periodical variations of the geomagnetic field, which are connected with the position of the Moon in relation to the Earth–Sun system. This phenomenon is explained by the fact that the trajectory in which the Earth's Moon moves in its orbit regularly crosses the trajectory of the solar wind movement. That is why there are geomagnetosphere parameters variations, which are correlated with the Moon cycle. In addition, although the Moon does not have its own obvious magnetosphere, there are some local magnetic anomalies near its surface. Their existence is caused by the ferromagnetic ilmenite in its soil, which has the same properties as a permanent magnet in conditions of iron oxide (Fe_2O_3) concentration of more than 25%.

27–Day Geomagnetic Variations The rotation of the Sun around its own axis causes the so–called *27–day geomagnetic variation.* It corresponds to the full rotation of the Sun around its own axis. The nature of these variations is concerned with the fact that the most active areas of the Sun throw solar plasma away to outer space not radially, but directionally. So, if these areas keep their activity during several solar rotations, there is a ~27−day frequency of geomagnetic variations.

Season Variations of the Geomagnetic Field The Earth's rotation around the Sun causes the so−called *season variations of the geomagnetic field.* These variations are connected with the mutual arrangement of terrestrial and solar equator planes. When these planes match each other (during the spring and fall equinoxes), there are some maximums in seasonal variations. In the case of the summer and winter solstices, there are minimums in seasonal variations. The reason for this fact is the increased concentration of active areas on the Sun. These areas are grouped in the zones of the 10°–30° north and south heliographic latitudes.

11−Year Variations of the Geomagnetic Field There are *11−year variations of the geomagnetic field*, which, due to their nature, are connected with variations in the Wolf number—a value that measures the number of sunspots. Based on recent research results, it has been found that this cycle period is not 11 years, but 22 years (two 11−year cycles of magnetic intensity vectors with reverse polarity). However, it is necessary to mention that the 11−year cycle is an average value of the geomagnetic variation period, as there are some known processes of similar origin with periods of 8–14 years.

Secular Variations of the Geomagnetic Field The substance motion and wave processes in the liquid conducting core of the Earth cause the so−called *secular variations of the geomagnetic field.* In many observations, the actual period of such geomagnetic variations is much less than 100 years and it is usually limited to the range of 10–20 years. In addition, research and analysis of the secular variations of the geomagnetic field provide objective information and knowledge about the geomagnetosphere variations. Nowadays, this is one of the main approaches for studying the processes occurring inside the planet's core.

Magnetic Storms The *solar wind* is a flow of ionized particles, which are radially catapulted from the solar corona to outer space. This natural phenomenon causes *magnetic storms*. The magnetic storms are strong perturbations of the Earth's magnetic field because of the solar flares, which sharply distort its motion relative to the slow periodic variations. The average duration of these storms is usually several tens of hours.

The magnetic storm is one of the main variants of the most common geophysical process—the magnetosphere storm. It is followed by auroras in the upper atmosphere of the Earth, ionospheric disturbances, X−ray radiation and low−frequency radiation. During the magnetic storms, there is a significant change of the ionosphere layer parameters, which reflect and absorb radio waves (their height, electron density and other parameters).

Magnetic storms lead to significant interferences in shortwave radio communications. The magnetic disturbances are responsible for the heat of the upper atmosphere and the transfer of heat down to the troposphere, which causes cyclones and circulations inside the troposphere.

Exospace Magnetic Activity *Exospace magnetic activity* describes a complex of geomagnetic disturbances caused by the increased magnetic activities of celestial bodies (apart from the Sun) both inside the solar system (Jupiter, Saturn, Uranus, Neptune, comets, etc.) and outer space. For example, the magnetic field of Jupiter is considered to be ten times stronger than Earth's, and its magnetic moment is about 18,000 times larger (Russell 1993). This fact influences the parameters and characteristics of the geomagnetic field.

Schumann Resonance *Schumann resonance* is a phenomenon of forming standing electromagnetic waves with extremely low, low and super–low frequencies transmitted between the Earth's surface and the ionosphere. The origin of the phenomenon is a number of lightning flows, which emit electromagnetic energy at frequencies of up to 100 kHz to outer space (Schlegel 2002).

Geomagnetic Anomalies *Geomagnetic anomalies* describe intraterrestrial natural sources of distortion of the magnetic intensity vector. They are caused by magnetization of the top parts of the crust. In the majority of cases, geomagnetic activity distorts the geomagnetic field by no more than 200 μT. Nevertheless, there are some well–known exceptions, for example the *Kursk magnetic anomaly*, the *Eastern Siberian magnetic anomaly*, and the *South Atlantic magnetic anomaly*. All these anomalies can cause variations of units and tens of μT.

According to information published by the Russian National Aeronautics and Space Administration, about 2% of the Earth's surface requires further investigation for *geomagnetic anomalies*. This 2% includes 36 countries, such as Qatar, Yemen, Pakistan, Indonesia, Jamaica, Cambodia, Laos, Bahrain, Kuwait, Guinea, Fiji, Gabon, Honduras, Chile, Nepal, and Myanmar.

Technogenic Factors In big cities and industrial centers, *technogenic factors* are dominant causes of local distortions of geomagnetic field parameters. Unlike geomagnetic variations, which are caused by natural events and processes, the frequency spectrum of technogenic–originated magnetic variations is quite wide and includes the frequencies from 0 Hz to radio frequencies of microwaves. Special attention should be paid to the frequencies of the industrial spectrum. They belong to the range of super–low frequencies and are widespread in both industrial and residential areas. So, the range of industrial frequencies in Russia and Europe is defined as 50 Hz. In some other countries, this type of frequency is equal to 60 Hz.

The main sources of electromagnetic fields of industrial frequency are various types of distributors, and consumers and generators of electricity (electrical substations, electric power lines of high voltage, electric vehicles, thermal power stations, hydroelectricity, nuclear power plants, tidal power, various alternative generators of electricity, and others) (Vorobev and Garipova 2012).

The field characteristics of technogenic–originated geomagnetic variations are very ambiguous and depend on the power of engines, the level of their magnetic shielding, the distance between the magnetic disturbance source and its observation location, and the presence of massive ferromagnetic objects within the close area.

The events and processes that cause geomagnetic variations have various and often independent origins. Local/global geomagnetosphere deformations are very different in both amplitude frequency and probabilistic estimations and characteristics. As a result, the actual visualization of the geomagnetic field is a complicated superposition of a probabilistic set of geomagnetic variations, which are caused by a number of various events. This is the reason why their estimation using traditional physical quantities is not so accurate.

In this regard, the study of events that are connected with geomagnetic variations is often related to the term *geomagnetic activity*. It defines a degree of geomagnetic field disturbances for the appropriate time period during the interaction of the Sun's corpuscular radiation with the Earth's magnetosphere.

V. I. Vernadsky and A. L. Chizhevsky were among the first scientists who paid great attention to the problem of solar–Earth relations. Besides introducing the hypothesis about biosphere interaction with natural geomagnetic variations, they also introduced and pointed out the correlation of biosphere members' mortality with the Sun's specific radiations (Chizhevsky 1976; Vernadsky 2004).

Nowadays, researchers and specialists in biology, medicine, geophysics, geology, technics, sociology, psychology, and many other sciences (Table 3.1) pay great attention to correlations between external geomagnetic variations and the evolution of various objects and systems, and analyze such findings in their own sciences.

This interest is based on the idea and proof that (according to known statistics) some components of geomagnetic variations or their combinations influence biological, technical, geological, and other objects and systems, especially living organisms. It has been observed that distortions in the normal conditions of existence force objects and systems to either adapt to magnetic state changes (via deformation, mutation, etc.) or continue existing there in a stressed (unstable) mode.

3.4 Mathematical Modeling of the Geomagnetic Field

The full vector of the Earth's magnetic field intensity at any geographical point with spatiotemporal coordinates (latitude, longitude, altitude, year) is defined as the sum of three components (Afanasiev et al. 1968) (Eq. 3.1):

$$\boldsymbol{B}_{\mathrm{ge}} = \boldsymbol{B}_1 + \boldsymbol{B}_2 + \boldsymbol{B}_3, \tag{3.1}$$

where $\boldsymbol{B}_1$ is an intensity vector of the geomagnetic field of intraterrestrial sources; $\boldsymbol{B}_2$ is a regular component of the intensity vector of the geomagnetic field of magnetosphere currents, which is calculated in the solar magnetosphere coordinate system; and $\boldsymbol{B}_3$ is a geomagnetic field intensity vector component with a technogenic origin.

Table 3.1 Human activity spheres dependent on geomagnetic variation parameters

Technics				Earth sciences					Living objects and systems					
Navigation	Aviation, astronautics	Mining	Metrology	Meteorology	Geology	Geophysics	Seismology	Volcanology	Biology and medicine	Ecology	Agriculture	Sociology	Psychology	Paleontology

The *magnetic field of intraterrestrial sources* consists of two parts: the electric current field in the core (the main field), which is ~98% of the whole field, and the magnetism field of rocks, which is ~2% of the whole field. Thus, with a rise in altitude the crust field decreases much more quickly than the main field, and from an altitude of ~100 km its value can be ignored (Standard GOST 25645.127–85 1985c; Standard GOST 25645.126–85 1985b).

The model of the *magnetosphere current magnetic field* defines a regular component of the geomagnetosphere. This model depends on parameters of the interplanetary medium. It demonstrates the well–known compression of the Earth's magnetosphere on the sunlit side due to interaction with the solar wind, day–night asymmetry (the field on the night side of the Earth is weakened), and diurnal, seasonal, yearly, and other variations of the geomagnetic field.

The summand $\boldsymbol{B}_3$ represents the magnetic field component of technogenic origin, which occurs due to human activity and usually has wide and unpredictable amplitudes and frequency ranges.

The normal geomagnetic field is estimated as a value of the $\boldsymbol{B}_1$ vector with exclusion of a component, which is caused by the magnetic properties of rocks (including magnetic anomalies). This component is excluded as a geomagnetic variation in Eq. 3.2:

$$\boldsymbol{B}_0 = \boldsymbol{B}_1 - \Delta \boldsymbol{B}_1', \tag{3.2}$$

where $\boldsymbol{B}_0$ is the undisturbed geomagnetic field intensity at the point with spatio-temporal coordinates; and $\Delta \boldsymbol{B}'_1$ is the component of intraterrestrial sources of the geomagnetic field, which represents the magnetic properties of rocks.

So, Eq. 3.3 defines geomagnetic variations at the point with spatiotemporal coordinates:

$$\boldsymbol{B}_{\mathrm{GMV}} = \Delta \boldsymbol{B}_1' + \boldsymbol{B}_2 + \boldsymbol{B}_3, \tag{3.3}$$

So, the approach to geomagnetic field registration with the magnetic activity indices M and σM is provided by Eqs. 3.4 and 3.5:

$$M = \boldsymbol{B}_{\mathrm{ge}} - \boldsymbol{B}_0 \quad \text{or} \quad M = \Delta \boldsymbol{B}_1' + \boldsymbol{B}_2 + \boldsymbol{B}_3; \tag{3.4}$$

$$\sigma M = \frac{\boldsymbol{B}_{\mathrm{ge}} - \boldsymbol{B}_0}{\boldsymbol{B}_{\mathrm{ge}}} \cdot 100\% \quad \text{or} \quad \sigma M = \frac{\Delta \boldsymbol{B}_1' + \boldsymbol{B}_2 + \boldsymbol{B}_3}{\boldsymbol{B}_1 + \boldsymbol{B}_2 + \boldsymbol{B}_3} \cdot 100\%, \tag{3.5}$$

where $\boldsymbol{B}_{ge}$ is an empirical value of geomagnetic field intensity; and $\boldsymbol{B}_0$ is an analytical value of geomagnetic field intensity.

High magnetization is caused by the presence of such ferromagnetic minerals as magnetite ($FeO \bullet Fe_2O_3$), titano–magnetite ($FeTiO_3 \bullet Fe_3O_4$), hematite (Fe_2O_3), and pyrrhotite (FeS). The magnetic susceptibility χ of these minerals is defined on the basis of various fields, with supposition of a wide range, and is described by Eq. 3.6:

$$\chi = \frac{\boldsymbol{I}}{\boldsymbol{H}}, \tag{3.6}$$

where $\boldsymbol{I}$ is the magnetization of rocks; and $\boldsymbol{H}$ is the intensity of the geomagnetic field.

So, a parameter χ is one of the proportionality multipliers in calculating the intensity of the magnetization of rocks in the geomagnetic field.

The presence of magnetite (or other ferromagnetic minerals) in the rock has an influence on the χ parameter value. But the χ value depends not just on the quantity of the magnetite in the rock, but also on the rock's structural and textural features. The research has demonstrated that there is no proportional connection between the presence of magnetite and the χ value. This fact and the irregular content of ferromagnetic minerals in similarly named rocks explain a wide range of χ changes in many types of rocks.

Large changes in the χ value require mass probes of the same types of rocks to get a reliable value of magnetic susceptibility. Values of χ are used for preliminary estimation of magnetic field changes and in geological explanation of magnetic anomalies.

Estimation of the expected anomaly intensity is based on Eq. (3.7, which refers to the case where the visibility angle of the top edge from the observation point is close to 180°:

$$\boldsymbol{Z} = (\chi - \chi_0)\boldsymbol{H}, \tag{3.7}$$

where the value χ_0 corresponds to the adjoining rocks, i.e., the rocks that contain an ore body, vein, or other geological object with a mineral resource. (In inclined bedding of the objects, the adjoining rock is called lateral rock).

It has been defined that the majority of rocks have residual magnetization $\boldsymbol{I}_r$, whose value is occasionally many times greater than induced $\boldsymbol{I}_i$ (especially in the case of ferruginous quartzites or peridotites).

In this case, the intensity of magnetic anomalies mainly depends on the $\boldsymbol{I}_r$ value. However, in general, the equations for calculating the intensity of anomalies must be included in Eq. (3.7).

3.4.1 Mathematical Modeling of the Normal Geomagnetic Field

In order to solve the problem of analytical estimation of the $\boldsymbol{B}_0$ parameter, it is helpful firstly to represent the main field model based on a spherical harmonic series, which depends on geographical coordinates.

The scalar potential of intraterrestrial sources of geomagnetic field induction U [nT · km] at any point with spherical coordinates r, θ, and λ is defined by Eq. 3.8:

$$U = R_E \sum_{n=1}^{N} \sum_{m=0}^{n} \left(g_n^m \cos(m\lambda) + h_n^m \sin(m\lambda) \right) \left(\frac{R_E}{r} \right)^{n+1} P_n^m \cos\theta, \tag{3.8}$$

where r is the distance from the Earth's center to the observation point (geocentric distance) [km]; λ is the longitude from the Greenwich meridian [degrees]; θ is the polar angle (colatitude, $\theta = (\pi/2) - \varphi'$ [degrees], where φ' is the latitude in spherical coordinates [degrees]); R_E is the average radius of the Earth, $R_E = 6371.03$ km; $g_n{}^m(t)$,$h_n{}^m(t)$ are spherical harmonic coefficients [nT], which depend on the time period; and $P_n{}^m$ are Schmidt semi−normalized associated Legendre functions of degree n and order m (Fanselan et al. 1964).

In the geophysical literature, Eq. 3.8 is widely known as a Gaussian, and it is generally recognized as an international standard for the undisturbed geomagnetic field.

There has been a great amount of spherical harmonic analysis. However, the problem of spherical harmonic optimal length definition is still acute (Yanovsky 1978).

Values of higher orders for elements $g_n{}^m$ and $h_n{}^m$ were calculated by G. Phanzelau (up to the 15th order) and by V. Kolosov and E. Kropachev (up to the 23rd order). The main conclusion about these values is that there is a sharp decrease of all coefficients for 6–8 orders and then a slow decrease with some oscillations. There are no coefficients (up to the 23rd order) with a sharp increase, which could point to magnetic field local sources.

Thus, the analyses of great amounts of elements prove the hypothesis of Gauss about convergence of spherical harmonics, which represents geomagnetic potential. As usual in spherical harmonic analyses, the harmonics are limited by 8–10 elements. But, for sufficiently homogeneous and highly accurate data (for example, as in satellite imaging), the harmonics series can be extended up to 12 and 13 harmonics (Panteleev 2000).

For example, for series of 11–13 harmonics the error in geomagnetic field calculation on the Earth's surface is less than ~2%. In other words, with values of coefficients $g_n{}^m$ and $h_n{}^m$ up to the 12th order and the definition of an undisturbed geomagnetic field, Eq. 3.9 is defined:

$$\boldsymbol{B}_0 = \boldsymbol{B}_1 - \Delta \boldsymbol{B}_1' \pm 2\%. \tag{3.9}$$

Due to the main field temporal variations, the coefficients of the harmonic series (spherical harmonic coefficients) are periodically (once every 5 years) recalculated with the new experimental data.

The change in the main field for 1 year (or secular variation) is also represented by spherical harmonics series, which are available at http://www.ngdc.noaa.gov/IAGA/vmod/igrf11coeffs.txt.

The Schmidt semi−normalized associated Legendre functions $P_n{}^m$ from Eq. 3.9 in general can be defined as an orthogonal polynomial, which is represented by Eq. 3.10 (Panteleev 2000; Kampe de Ferrier 1963):

$$P_n^m(\cos(\theta)) = 1\cdot 3\cdot 5\ldots\sqrt{\frac{\varepsilon_m}{(n+m)!(n-m)!}} \times \sin^m\theta\left[\cos^{n-m}\theta - \frac{(n-m)(n-m-1)}{2(2n-1)}\cos^{n-m-2} + \frac{(n-m)(n-m-1)(n-m-2)(n-m-3)}{2\cdot 4(2n-1)(2n-3)}\cos^{n-m-4}\theta - \ldots\right], \tag{3.10}$$

where ε_m is a normalization factor ($\varepsilon_m = 2$ for $m \geq 1$ and $\varepsilon_m = 1$ for $m = 0$); n is a degree of spherical harmonics; and m is an order of spherical harmonics.

In a number of scientific problems, some geospatial data (for example, positions of satellites in space) are represented by coordinates φ, λ, h, which are based on the Earth's surface approximation as a spheroid. In these problems, the Earth's ellipticity is neglected with no difference between spherical and geodesic coordinates. But in accurate calculations, the Earth's pole compression should be taken into account.

Thereby, the values r and φ' with the compression taken into account are defined by Eqs. 3.11 and 3.12:

$$r^2 = h^2 + 2h\sqrt{a^2\cos^2\varphi + b^2\sin^2\varphi} + \frac{a^4\cos^2\varphi + b^4\sin^2\varphi}{a^2\cos^2\varphi + b^2\sin^2\varphi}; \tag{3.11}$$

$$\mathrm{tg}\ \varphi' = \frac{b^2 + h\sqrt{a^2\cos^2\varphi + b^2\sin^2\varphi}}{a^2 + h\sqrt{a^2\cos^2\varphi + b^2\sin^2\varphi}}\mathrm{tg}\ \varphi, \tag{3.12}$$

where φ is the geographical (geodesic) latitude of the point [degrees]; h is the altitude (elevation) [degrees]; a is the semi−major axis of Earth's ellipsoid; and b is the semi−minor axis of Earth's ellipsoid.

The longitudes λ in spherical and geodesic coordinates are identical. So, the components X', Y', and Z' of the induction vector of the intraterrestrially sourced geomagnetic field in nanoteslas are defined by Eqs. 3.13, 3.14 and 3.15:

$$X' = \frac{1}{r}\frac{\partial U}{\partial \theta} = \sum_{n=1}^{N}\sum_{m=0}^{n}\left(g_n^m\cos(m\lambda) + h_n^m\sin(m\lambda)\right)\left(\frac{R_E}{r}\right)^{n+2}\frac{\partial P_n^m\cos(\theta)}{\partial\theta}; \tag{3.13}$$

$$\begin{aligned}\boldsymbol{Y}' &= -\frac{1/r}{\sin\theta}\frac{\partial U}{\partial\theta}\\ &= \sum_{n=1}^{N}\sum_{m=0}^{n} m\left(g_n^m \sin(m\lambda) - h_n^m \cos(m\lambda)\right)\left(\frac{R_E}{r}\right)^{n+2}\frac{P_n^m \cos(\theta)}{\sin\theta};\end{aligned} \qquad (3.14)$$

$$\begin{aligned}\boldsymbol{Z}' &= \frac{\partial U}{\partial r}\\ &= -\sum_{n=1}^{N}\sum_{m=0}^{n}(n+1)\left(g_n^m \cos(m\lambda) + h_n^m \sin(m\lambda)\right)\left(\frac{R_E}{r}\right)^{n+2} P_n^m \cos(\theta);\end{aligned} \qquad (3.15)$$

Thus, the geomagnetic field induction vector distribution in space is described by a number of geomagnetic elements:

- Orthogonal components of the geomagnetic field
- The modulus of the geomagnetic field induction vector
- Angle elements of the geomagnetic field (geomagnetic dipole)

Thus, in the point A (φ, λ, h) the orthogonal components of the induction vector (in geodesic coordinate systems) are defined by Eqs. 3.16, 3.17, 3.18, and 3.19, and the module of the geomagnetic field induction vector is defined by Eq. 3.23:

$$\boldsymbol{X} = \boldsymbol{X}'\cos(\varphi - \varphi') + \boldsymbol{Z}'\sin(\varphi - \varphi'); \qquad (3.16)$$

$$\mathbf{Y} = \mathbf{Y}'; \qquad (3.17)$$

The values of field element $\boldsymbol{Y}$ for the point with $\theta = 0$ are calculated by linear interpolation:

$$\boldsymbol{Z} = \boldsymbol{Z}'\cos(\varphi - \varphi') + \boldsymbol{X}'\sin(\varphi - \varphi'); \qquad (3.18)$$

$$|\boldsymbol{H}| = \sqrt{\boldsymbol{X}^2 + \boldsymbol{Y}^2}, \qquad (3.19)$$

where $|\boldsymbol{H}|$ is a horizontal component of the geomagnetic field induction vector $\boldsymbol{B}_0$ ($\boldsymbol{B}_0$ projection on the horizontal plane XY).

$$|\boldsymbol{B}_0| = \sqrt{\boldsymbol{X}^2 + \boldsymbol{Y}^2 + \boldsymbol{Z}^2}, \qquad (3.20)$$

Angle elements of the geomagnetic dipole are defined by Eq. 3.21:

$$D = \mathrm{arctg}(\boldsymbol{Y}/\boldsymbol{X}); \quad I = \arcsin(\boldsymbol{Z}/|\boldsymbol{B}_0|), \qquad (3.21)$$

where D is the magnetic declination—the angle between the geographic and magnetic meridians (positive to the east); and I is the magnetic dip—the angle between the horizontal plane XY and a vector $\boldsymbol{B}_0$ (positive to the east).

Geomagnetic elements for the appropriate year t are calculated with spherical harmonic coefficients $g_n^m(t)$ and $h_n^m(t)$, which are recalculated (actualized) according to Eqs. 3.22 and 3.23:

$$g_{n,t}^m = g_{n,t0}^m + \underline{g_n^m}(t - t_0); \tag{3.22}$$

$$h_{n,t}^m = h_{n,t0}^m + \underline{h_n^m}(t - t_0); \tag{3.23}$$

where t is the current year; t_0 is the year of the known spherical harmonic coefficients $g_n^m(t)$, $h_n^m(t)$; and $\underline{g}_n^m$ and $\underline{h}_n^m$ are the spherical harmonic coefficient amendments, which are connected with secular variations of the geomagnetic field.

The dipole geomagnetic field (Fig. 3.1) corresponds to the field represented by the first element of spherical harmonics. The components of a dipole element are defined by Eqs. 3.24, 3.25, and 3.26:

$$\boldsymbol{X}(r,\theta,\lambda) = \left[-g_1^0 \sin\theta + \left(g_1^1 \cos\lambda + h_1^1 \sin\lambda\right)\cos\theta\right]\left(\frac{R_E}{r}\right)^3; \tag{3.24}$$

$$\boldsymbol{Y}(r,\theta,\lambda) = \left[g_1^1 \sin\lambda - h_1^1 \cos\lambda\right]\left(\frac{R_E}{r}\right)^3; \tag{3.25}$$

$$\boldsymbol{Z}(r,\theta,\lambda) = -2\left[g_1^0 \cos\theta + \left(g_1^1 \cos\lambda + h_1^1 \sin\lambda\right)\sin\theta\right]\left(\frac{R_E}{r}\right)^3. \tag{3.26}$$

The coordinates of the magnetic dipole poles (geomagnetic poles) and the magnetic moment M [T·m^3] are defined by Eqs. 3.27, 3.28, and 3.29:

$$\mathrm{tg}\ \Phi_0 = \frac{g_1^0}{\sqrt{\left(g_1^1\right)^2 + \left(h_1^1\right)^2}}; \tag{3.27}$$

$$\mathrm{tg}\ \Lambda_0 = \frac{h_1^1}{g_1^1}; \tag{3.28}$$

$$M = R_E^3\sqrt{\left(g_1^0\right)^2 + \left(g_1^1\right)^2 + \left(h_1^1\right)^2}. \tag{3.29}$$

where Φ_0 and Λ_0 are the geographical latitude and longitude of the geomagnetic pole [degrees].

3.4.2 Method and Example of Normal Geomagnetic Field Parameter Calculation

Let us assume that there is a geographical point $\{A\}$ with spatiotemporal coordinates: $h = 0.2$ km; $\varphi = 54.7384°$N; $\lambda = 55.9832°$E; date: June 30, 2014.

The parameters of the geomagnetic field at the point $\{A\}$ are suggested to be calculated according to the method represented in Fig. 3.3 (Vorobev 2013a, b, c), and the parameters of the geomagnetic field are calculated as a result of stages 3 and 7–9 of the proposed method.

The proposed method suggests that in order to calculate the parameters of the undisturbed geomagnetic field, it is necessary and sufficient to define spatiotemporal coordinates, geometric parameters of the Earth and a set of spherical harmonic coefficients. These sets of spherical harmonic coefficients are represented as square matrices of size $[(n_{max} + 1) \times (m_{max} + 1)]$, where n is a degree of spherical harmonics; and m is an order of spherical harmonics (Tables 3.2, 3.3, and 3.4).

According to the method presented in Fig. 3.3, the matrices in Tables 3.2, 3.3, and 3.4 with the coefficients are recalculated with amendment for geomagnetic field secular variation. For the particular date of the problem, the year is calculated as a repeating decimal (for example, the year of June 30, 2014, is written as "2014.5"). The result is represented as a set of actualized matrices of spherical harmonic

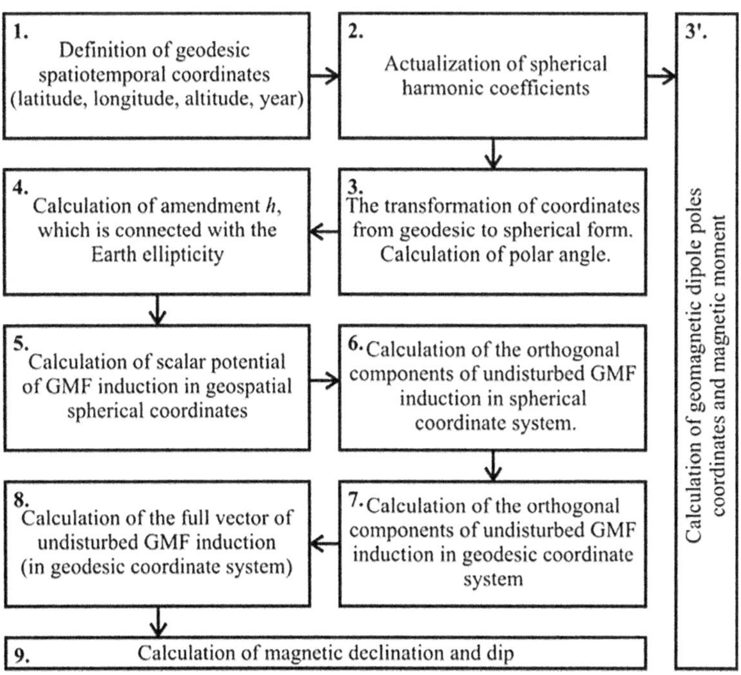

Fig. 3.3 Method of undisturbed geomagnetic field parameter calculation

Table 3.2 Matrix of coefficients $g^{m}_{n,t0}$

0	0	0	0	0	0	0	0	0	0	0	0	0
−29,497	−1585.9	0	0	0	0	0	0	0	0	0	0	0
−2396.6	3026.0	1668.6	0	0	0	0	0	0	0	0	0	0
1339.7	−2326.3	1231.7	634.2	0	0	0	0	0	0	0	0	0
912.6	809.0	166.6	−357.1	89.7	0	0	0	0	0	0	0	0
−231.1	357.2	200.3	−141.2	−163.1	−7.7	0	0	0	0	0	0	0
72.8	68.6	76.0	−141.4	−22.9	13.1	−77.9	0	0	0	0	0	0
80.4	−75.0	−4.7	45.3	14.0	10.4	1.6	4.9	0	0	0	0	0
24.3	8.2	−14.5	−5.7	−19.3	11.6	10.9	−14.1	−3.7	0	0	0	0
5.4	9.4	3.4	−5.3	3.1	−12.4	−0.8	8.4	−8.4	−10.1	0	0	0
−2.0	−6.3	0.9	−1.1	−0.2	2.5	−0.3	2.2	3.1	−1.0	−2.8	0	0
3.0	−1.5	−2.1	1.6	−0.5	0.5	−0.8	0.4	1.8	0.2	0.8	3.8	0
−2.1	−0.2	0.3	1.0	−0.7	0.9	−0.1	0.5	−0.4	−0.4	0.2	−0.8	0

Epoch (2010)

Table 3.3 Matrix of coefficients $h^{m}_{n,t0}$

0	0	0	0	0	0	0	0	0	0	0	0	0
0	4945.1	0	0	0	0	0	0	0	0	0	0	0
0	−2707.7	−575.4	0	0	0	0	0	0	0	0	0	0
0	−160.5	251.7	−536.8	0	0	0	0	0	0	0	0	0
0	286.4	−211.2	164.4	−309.2	0	0	0	0	0	0	0	0
0	44.7	188.9	−118.1	0.1	100.9	0	0	0	0	0	0	0
0	−20.8	44.2	61.5	−66.3	3.1	54.9	0	0	0	0	0	0
0	−57.8	−21.2	6.6	24.9	7.0	−27.7	−3.4	0	0	0	0	0
0	10.9	−20.0	11.9	−17.4	16.7	7.1	−10.8	1.7	0	0	0	0
0	−20.5	11.6	12.8	−7.2	−7.4	8.0	2.2	−6.1	7.0	0	0	0
0	2.8	−0.1	4.7	4.4	−7.2	−1.0	−4.0	−2.0	−2.0	−8.3	0	0
0	0.1	1.7	−0.6	−1.8	0.9	−0.4	−2.5	−1.3	−2.1	−1.9	−1.8	0
0	−0.8	0.3	2.2	−2.5	0.5	0.6	0.0	0.1	0.3	−0.9	−0.2	0.8

Epoch (2010)

Table 3.4 Matrix of coefficients $\underline{g}^m{}_n$

0	0	0	0	0	0	0	0	0	0	0	0	0
11.4	16.7	0	0	0	0	0	0	0	0	0	0	0
−11.3	−3.9	2.7	0	0	0	0	0	0	0	0	0	0
1.3	−3.9	−2.9	−8.1	0	0	0	0	0	0	0	0	0
−1.4	2.0	−8.9	4.4	−2.3	0	0	0	0	0	0	0	0
−0.5	0.5	−1.5	−0.7	1.3	1.4	0	0	0	0	0	0	0
−0.3	−0.3	−0.3	1.9	−1.6	−0.2	1.8	0	0	0	0	0	0
0.2	−0.1	−0.6	1.4	0.3	0.1	−0.8	0.4	0	0	0	0	0
−0.1	0.1	−0.5	0.3	−0.3	0.3	0.2	−0.5	0.2	0	0	0	0
0	0	0	0	0	0	0	0	0	0	0	0	0
0	0	0	0	0	0	0	0	0	0	0	0	0
0	0	0	0	0	0	0	0	0	0	0	0	0
0	0	0	0	0	0	0	0	0	0	0	0	0

Epoch (2010)

Table 3.5 Matrix of coefficients $\underline{h}^m{}_n$

0	0	0	0	0	0	0	0	0	0	0	0	0
0	−28.8	0	0	0	0	0	0	0	0	0	0	0
0	−23.0	−12.9	0	0	0	0	0	0	0	0	0	0
0	8.6	−2.9	−2.1	0	0	0	0	0	0	0	0	0
0	0.4	3.2	3.6	−0.8	0	0	0	0	0	0	0	0
0	0.5	1.5	0.9	3.7	−0.6	0	0	0	0	0	0	0
0	−0.1	−2.1	−0.4	−0.5	0.8	0.5	0	0	0	0	0	0
0	0.6	0.3	−0.2	−0.1	−0.8	−0.3	0.2	0	0	0	0	0
0	0.0	0.2	0.5	0.4	0.1	−0.1	0.4	0.4	0	0	0	0
0	0	0	0	0	0	0	0	0	0	0	0	0
0	0	0	0	0	0	0	0	0	0	0	0	0
0	0	0	0	0	0	0	0	0	0	0	0	0
0	0	0	0	0	0	0	0	0	0	0	0	0

Epoch (2010)

coefficients (Tables 3.4 and 3.5 for the year 2014.5), which are calculated according to Eqs. 3.30 and 3.31 (Tables 3.6 and 3.7):

$$g_t[n,m] = g_{t0}[n,m] + \underline{g}_t[n,m] \cdot (2014.5 - 2010). \tag{3.30}$$

$$h_t[n,m] = h_{t0}[n,m] + \underline{h}_t[n,m] \cdot (2014.5 - 2010). \tag{3.31}$$

Latitude in spherical coordinates is calculated using Eq. 3.32:

$$\phi' = \mathrm{arctg}\left(\frac{b^2 + h\sqrt{a^2\cos^2\phi + b^2\sin^2\phi}}{a^2 + h\sqrt{a^2\cos^2\phi + b^2\sin^2\phi}}\,\mathrm{tg}\,\phi\right) = 0.95220\ \mathrm{rad}$$
$$= 54.557^\circ; \tag{3.32}$$

The polar angle is calculated using Eq. 3.33:

Table 3.6 Matrix of coefficients $g^{m}_{n,t}$ with amendment for secular variation

0	0	0	0	0	0	0	0	0	0	0	0	0
−29,445	−1510.8	0	0	0	0	0	0	0	0	0	0	0
−2447.5	3008.45	1680.75	0	0	0	0	0	0	0	0	0	0
1345.55	−2343.9	1218.65	597.75	0	0	0	0	0	0	0	0	0
906.3	818	126.55	−337.3	79.35	0	0	0	0	0	0	0	0
−233.35	359.45	193.55	−144.4	−157.3	−1.4	0	0	0	0	0	0	0
71.45	67.25	74.65	−132.9	−30.1	12.2	−69.8	0	0	0	0	0	0
81.3	−75.45	−7.4	51.6	15.35	10.85	−2	6.7	0	0	0	0	0
23.85	8.65	−16.75	−4.35	−20.65	12.95	11.8	−16.35	−2.8	0	0	0	0
5.4	9.4	3.4	−5.3	3.1	−12.4	−0.8	8.4	−8.4	−10.1	0	0	0
−2	−6.3	0.9	−1.1	−0.2	2.5	−0.3	2.2	3.1	−1	−2.8	0	0
3	−1.5	−2.1	1.6	−0.5	0.5	−0.8	0.4	1.8	0.2	0.8	3.8	0
−2.1	−0.2	0.3	1	−0.7	0.9	−0.1	0.5	−0.4	−0.4	0.2	−0.8	0

Table 3.7 Matrix of coefficients $h^m{}_{n,t}$ with amendment for secular variation

0	0	0	0	0	0	0	0	0	0	0	0	0
0	4815.5	0	0	0	0	0	0	0	0	0	0	0
0	−2811.2	−633.45	0	0	0	0	0	0	0	0	0	0
0	−121.8	238.65	−546.25	0	0	0	0	0	0	0	0	0
0	288.2	−196.8	180.6	−312.8	0	0	0	0	0	0	0	0
0	46.95	195.65	−114.05	16.75	98.2	0	0	0	0	0	0	0
0	−21.25	34.75	59.7	−68.55	6.7	57.15	0	0	0	0	0	0
0	−55.1	−19.85	5.7	24.45	3.4	−29.05	−2.5	0	0	0	0	0
0	10.9	−19.1	14.15	−15.6	17.15	6.65	−9	3.5	0	0	0	0
0	−20.5	11.6	12.8	−7.2	−7.4	8	2.2	−6.1	7	0	0	0
0	2.8	−0.1	4.7	4.4	−7.2	−1	−4	−2	−2	−8.3	0	0
0	0.1	1.7	−0.6	−1.8	0.9	−0.4	−2.5	−1.3	−2.1	−1.9	−1.8	0
0	−0.8	0.3	2.2	−2.5	0.5	0.6	0	0.1	0.3	−0.9	−0.2	0.8

$$\theta = (\pi/2) - \varphi' = 0.6186 \ \text{rad} = 35.443^{\circ}. \tag{3.33}$$

It is necessary to make an amendment for the Earth's ellipticity. So, the scalar potential of geomagnetic field induction of intraterrestrial sources at the point $\{A\}$ is calculated by Eqs. 3.34 and 3.35:

$$\begin{aligned} r = & \sqrt{h^2 + 2h\sqrt{a^2\cos^2\varphi + b^2\sin^2\varphi} + \frac{a^4\cos^2\varphi + b^4\sin^2\varphi}{a^2\cos^2\varphi + b^2\sin^2\varphi}} \\ & \approx 6364.1 \ \text{km}, \end{aligned} \tag{3.34}$$

$$\begin{aligned} U_{\{A\}}(r,\theta,\lambda) = \ & R_E \sum_{n=1}^{N}\sum_{m=0}^{n}(g_t[n,m]\cos(m\lambda) + h_t[n,m]\sin(m\lambda))\left(\frac{R_E}{r}\right)^{n+1} \times \prod_{n=1}^{n}(2n-1) \\ & \times \sqrt{\frac{\varepsilon_m}{(n+m)!(n-m)!}}\sin^m\theta \cdot \Big[\cos^{n-m}\theta - \frac{(n-m)(n-m-1)}{2(2n-1)} \\ & \times \cos^{n-m-2}\theta + \frac{(n-m)(n-m-1)(n-m-2)(n-m-3)}{2\cdot 4(2n-1)(2n-3)}\cos^{n-m-4}\theta \\ & - \frac{(n-m)(n-m-1)(n-m-2)(n-m-3)(n-m-4)(n-m-5)}{2\cdot 4\cdot 6(2n-1)(2n-3)(2n-5)} \\ & \times \cos^{n-m-6}\theta + \frac{(n-m)(n-m-1)(n-m-2)(n-m-3)(n-m-4)}{2\cdot 4\cdot 6\cdot 8(2n-1)(2n-3)(2n-5)} \\ & \times \frac{(n-m-5)(n-m-6)(n-m-7)}{(2n-7)} \times \cos^{n-m-8}\theta \\ & - \frac{(n-m)(n-m-1)(n-m-2)(n-m-3)(n-m-4)(n-m-5)}{2\cdot 4\cdot 6\cdot 8\cdot 10(2n-1)(2n-3)(2n-5)} \\ & \times \frac{(n-m-6)(n-m-7)(n-m-8)(n-m-9)}{(2n-7)(2n-9)} \times \cos^{n-m-10}\theta \\ & + \frac{(n-m)(n-m-1)(n-m-2)(n-m-3)(n-m-4)(n-m-5)}{2\cdot 4\cdot 6\cdot 8\cdot 10\cdot 12(2n-1)(2n-3)(2n-5)(2n-7)} \\ & \times \frac{(n-m-6)(n-m-7)(n-m-8)(n-m-9)(n-m-10)}{(2n-9)} \\ & \times \frac{(n-m-11)}{(2n-11)}\cos^{n-m-12}\theta\Big] \approx -1.58885829 \cdot 10^8 \text{nT} \cdot \text{km} \end{aligned} \tag{3.35}$$

Geomagnetic field induction vector components are defined by Eqs. 3.36, 3.37, and 3.38:

$$\boldsymbol{X}' = \frac{1}{r}\frac{\partial U}{\partial \theta} \approx 16{,}242.954 \ \text{nT}; \tag{3.36}$$

$$\boldsymbol{Y}' = -\frac{1}{r \cdot \sin\theta}\frac{\partial U}{\partial \lambda} \approx 3820.016 \ \text{nT}; \tag{3.37}$$

$$\boldsymbol{Z}' = \frac{\partial U}{\partial r} \approx 52{,}257.445 \ \text{nT}. \tag{3.38}$$

Equations 3.39, 3.40 and 3.41 define the orthogonal components of geomagnetic field induction:

$$\boldsymbol{X} = \boldsymbol{X}' \cos(\varphi - \varphi') + \boldsymbol{Z}' \sin(\varphi - \varphi') \approx 16,408,513 \ \text{nT}; \tag{3.39}$$

$$\mathbf{Y} = \mathbf{Y}' \approx 3820.016 \ \text{nT}; \tag{3.40}$$

$$\boldsymbol{Z} = \boldsymbol{Z}' \cos(\varphi - \varphi') + \boldsymbol{X}' \sin(\varphi - \varphi') = 52,205.697 \ \text{nT}; \tag{3.41}$$

Equation 3.42 defines the scalar of the induction full vector of the undisturbed geomagnetic field, Eqs. 3.43 and 3.44 define magnetic declination and inclination, and Eqs. 3.45 and 3.46 define the poles' coordinates:

$$|\boldsymbol{B}_0| = |\boldsymbol{B}_1| = \sqrt{\boldsymbol{X}^2 + \boldsymbol{Y}^2 + \boldsymbol{Z}^2} \approx 54.856 \ \mu T; \tag{3.42}$$

$$D = \text{arctg}(\boldsymbol{Y}/\boldsymbol{X}) \approx 13.105^{\circ}; \tag{3.43}$$

$$I = \arcsin(\boldsymbol{Z}/|\boldsymbol{B}_0|) \approx 72.115^{\circ}, \tag{3.44}$$

$$\Phi_0 = \text{arctg}\frac{g_1^0}{\sqrt{\left(g_1^1\right)^2 + \left(h_1^1\right)^2}} \approx -80.274^{\circ}; \tag{3.45}$$

$$\Lambda_0 = \text{arctg}\frac{h_1^1}{g_1^1} \approx -72.582^{\circ}. \tag{3.46}$$

According to Eq. 3.47, the magnetic moment quantitative estimation is the following:

$$M = R_{\text{E}}^3 \sqrt{\left(g_1^0\right)^2 + \left(g_1^1\right)^2 + \left(h_1^1\right)^2} \approx 7.726 \cdot 10^{15} \ \text{T} \cdot \text{m}^3. \tag{3.47}$$

Here, a method was presented and defined with the corresponding calculations for the specific point $\{A\}$ with spatiotemporal coordinates, where the normal geomagnetic field can be suggested as a field with characteristics within the limits 54.856 μT ± 2% (54.856 ± 1.097 μT).

Any deviations of the geomagnetic field induction vector values, which are recorded empirically, from the analytically calculated values for the point $\{A\}$ can be considered as local (maybe global) geomagnetosphere disturbances with further analysis of their genesis of amplitude and frequency range.

3.4.3 Research and Analysis of the Undisturbed Geomagnetic Field Near the Earth's Surface and in Circumterrestrial Space

The required calculations and the involved instruments are quite complex, and a computer program, GEOmagnetic_v1.0 (Vorobev 2013c), has been developed, which has as an input array the parameter $h(\theta, \lambda)$ forming (with an appropriate resolution) an array of undisturbed geomagnetic field induction vector distribution on the Earth's surface, and it also takes into account specific topographical features. These data are used as a base to produce a map of isolines of undisturbed geomagnetic field induction distribution (Fig. 3.4) or its three–dimensional model (Fig. 3.5).

In the more common case, these arrays can be generated both for estimation of the global (planetary) geomagnetic state and for analysis of regional distribution of undisturbed geomagnetic field features.

Figures 3.4 and 3.5 illustrate and prove that on a physical object moving in circumterrestrial space, due to the anisotropy of the undisturbed geomagnetic field, there are some geomagnetic field influences. These influences are comparable by their power with strong geomagnetic variations (according to the values of the K–index of the "Moskva" observatory).

Certainly, the known and published results in geomagnetic field research represent a specific character with dynamics of geomagnetic field parameters that are changing at an Earth point with spatiotemporal coordinates. But it is still not clear how significant the geomagnetic variation influence is on the object within its space in the undisturbed anisotropic geomagnetic field with a non–zero angle and/or linear speed of the object.

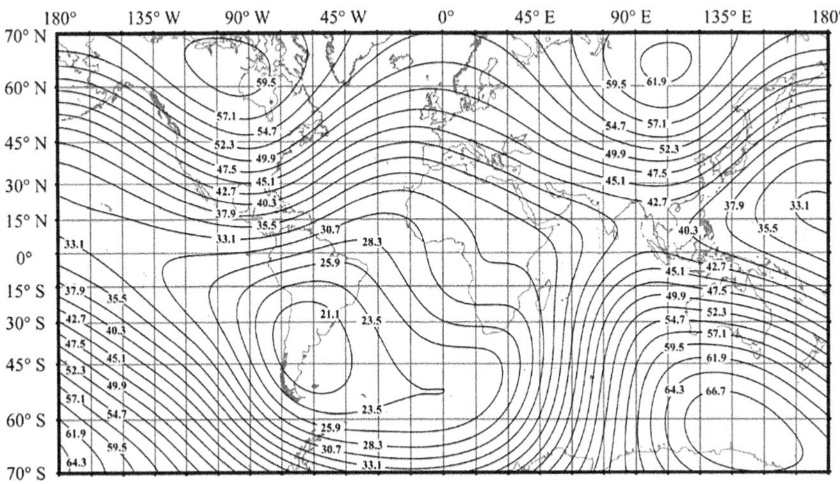

Fig. 3.4 Map of undisturbed geomagnetic field induction isolines

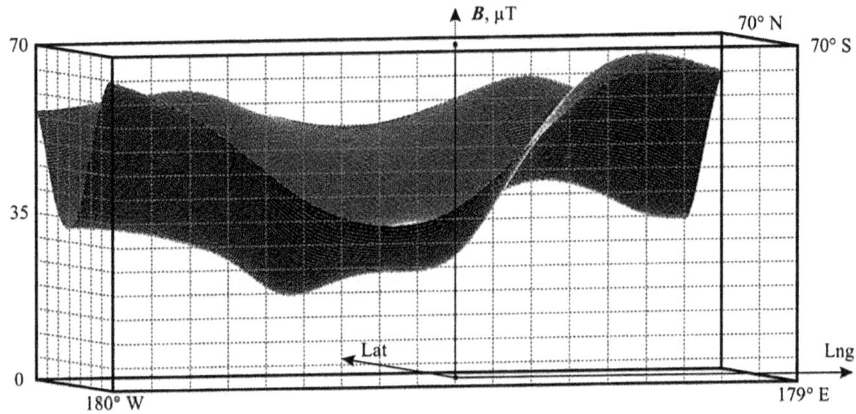

Fig. 3.5 Three–dimensional model of undisturbed geomagnetic field induction distribution

3.5 Mathematical Modeling of the Geomagnetic Pseudostorm Effect

The influence of the geomagnetic pseudostorm (GMPS) effect and the undisturbed geomagnetic field on a static ferromagnetic object can be compared with the influence of the density of anisotropic liquid flow, which changes direction and speed, on a static object. It is supposed that this analogy clearly represents the dynamics of real geomagnetic variation magnetic influences on the studied static object.

Using the same anisotropic properties of the liquid, it is useful to qualitatively estimate the influences on the object in conditions of a static environment and non–zero linear and/or angle speed. It is obvious that the dynamics of the influences are comparable with the dynamics from the previous example and depend on both the speed of the object and the gradient of environment inhomogeneity.

Furthermore, it would be useful to project this analogy onto a ferromagnetic object or system in an anisotropic magnetic field. Here, the term *geomagnetic pseudostorm* is introduced, which intends to represent the real geomagnetic field influence on the object in conditions of non–zero speed and undisturbed geomagnetic field anisotropy (Vorobev et al. 2013; Vorobev Geoinformatika 2013b).

The main parameters of the GMPS effect (Vorobev et al. 2014a, b, c; Vorobev and Shakirova 2014a, b; Vorobev patent for invention No. 2526234) are outlined below.

GMPS Range This is the difference between the maximal and minimal values of geomagnetic field induction in the area of the object, which is moving in an anisotropic magnetic field during the time period or at the distance:

$$\mathbf{B}_{\mathrm{GMPS}} = \mathbf{B}_{0\ \max} - \mathbf{B}_{0\ \min}, \tag{3.48}$$

where $\boldsymbol{B}_{0\ \mathrm{max}}$ and $\boldsymbol{B}_{0\ \mathrm{min}}$ are the maximal and minimal values of geomagnetic field induction [nT] in the area of the object, which is moving in an anisotropic magnetic field.

GMPS Frequency Spectrum This is a function of the distribution of the GMPS amplitude spectrum in the frequency area for continuous and discrete variants, which is defined by Eq. (3.49:

$$\boldsymbol{B}^*(f) = \int\limits_{-\infty}^{+\infty} \boldsymbol{B}_0(t)e^{-i2\pi ft}\mathrm{d}t \text{ or } \boldsymbol{B}^*(f) = \frac{1}{M}\sum_{t=0}^{M-1} \boldsymbol{B}_0(t)e^{\frac{-i2\pi ft}{M}}, \tag{3.49}$$

where $\boldsymbol{B}^*$ is the GMPS frequency spectrum; $\boldsymbol{B}_0$ is the value of the geomagnetic field induction in the location with given spatiotemporal coordinates; and M is the quantity of recorded values with constant discretization step by time.

Constant Component of GMPS This is a vector of harmonics superposition vertical shift, which represents the GMPS frequency spectrum:

$$\boldsymbol{B}_{//} = \frac{1}{M}\sum_{t=0}^{M-1} \boldsymbol{B}_0(t), \tag{3.50}$$

where M is the quantity of recorded values with the discretization step.

GMPS Intensity This is a physical quantity, which is numerically equal to the speed of the undisturbed geomagnetic field characteristic change in time relative to the frame of reference, which is connected to the moving object and depends on the object speed:

$$\boldsymbol{I}_{\mathrm{GMPS}} = \frac{\partial \boldsymbol{B}_0}{\partial t}, \tag{3.51}$$

where $\boldsymbol{I}_{\mathrm{GMPS}}$ is GMPS intensity [nT/s].

GMPS Potentiality This (geomagnetic induction gradient) is a vector, which is oriented to a three–dimensional space and points to the direction of the fastest increase of the undisturbed geomagnetic field induction absolute value. The vector by its absolute value is equal to the increase in the speed of $\boldsymbol{B}_0$ in the geographical direction [nT/rad; nT/rad; nT/km] and depends on the object position:

$$G_B = \nabla \boldsymbol{B}_0(\theta, \lambda, r) = \mathrm{grad}\ \boldsymbol{B}_0(\theta, \lambda, r) = \left(\frac{\partial \boldsymbol{B}_0}{\partial \theta}, \frac{\partial \boldsymbol{B}_0}{\partial \lambda}, \frac{\partial \boldsymbol{B}_0}{\partial r}\right), \tag{3.52}$$

where $\boldsymbol{B}_0$ is the induction (intensity) of the geomagnetic field in the point with spatiotemporal coordinates:

$$\begin{aligned} \boldsymbol{B}_0^2(\theta,\lambda,r)[\mathrm{nT}] &= \left[\frac{1}{r}\frac{\partial U}{\partial \theta}\cos(\varphi-\varphi') - \frac{\partial U}{\partial r}\sin(\varphi-\varphi')\right]^2 \\ &+ \left[-\frac{1}{r\cdot\sin\theta}\frac{\partial U}{\partial \lambda}\right]^2 + \left[-\frac{\partial U}{\partial r}\cos(\varphi-\varphi') - \frac{1}{r}\frac{\partial U}{\partial \theta}\sin(\varphi-\varphi')\right]^2. \end{aligned} \tag{3.53}$$

So, the analysis of geomagnetic field induction gradient distribution allows definition of an area of possible GMPS maximal intensity in the given geographical region. Therefore, the parameter G_B must be taken into account in developing aerospace navigation maps and flight paths. The geomagnetic isolines map (Fig. 3.5), GMPS definition, and typical geomagnetic field regions with geometrical centers in quadrilateral corners allow consideration that the maximal value of the geomagnetic induction gradient by latitude, and therefore GMPS intensity, while moving from north to south (from south to north), is located in the middle of the quadrilateral western edge, which coincides with the geometrical center of the Bermuda Triangle (Kush 1978). This situation can be explained by the fact that, due to the registered distribution of the geomagnetic isolines, the length of the quadrilateral western edge is minimal and therefore the intensity of isolines crossing its center is maximal.

Comparing this conclusion with the possible effects of GMPS influence on technical, biological, and other movable objects allows the "Bermuda Triangle" phenomenon to be proved scientifically. More research of GMPS effect parameters has demonstrated that the undisturbed geomagnetic field induction gradient by latitude and longitude has a wide range (0.5–50 μT/rad) and can be either positive or negative.

The geomagnetic field induction gradient by altitude is in range of $-9\ldots-30$ nT/km and always negative. This fact proves that the GMPS intensity within a biosphere activity area (the Earth's surface, lower layers of the troposphere) is always greater than in an area that is out of the limits of biosphere activity (higher layers of the troposphere, stratosphere, mesosphere, etc.)

GMPS effects must be taken into account by specialists in airspace instrumentation and scientists in medicine and biology who study the influences of weak nonionizing magnetic fields on living objects and systems.

Let us study the GMPS effect on the case of American Airlines flight route AA−973 from New York (JFK) to Rio de Janeiro (RIO). The approximate flight path of a Boeing 767−300, performing the flight route AA−973, is represented in Fig. 3.6 (Vorobev et al. 2014a, b).

The flight path is represented as an array of spatial coordinates which describe the airplane position taken during its flight in equal time intervals. The array allows calculation of geomagnetic field parameters for each set of spatial coordinates (Table 3.8).

The results of analysis of the data from Table 3.8 are presented for amplitude in Fig. 3.7a and frequency in Fig. 3.7b. Some interesting points in Fig. 3.7a are the $t1$–$t2$ takeoff time; $t2$–$t4$ flight at cruise speed at the altitude of 11033 m; $t4$–$t5$ landing time; and $t3$ pass over the equator.

Thereby, comparing and analyzing the values from Fig. 3.7 with the data in Table 3.8 allows us to draft the conclusion that the GMPS effect in this example

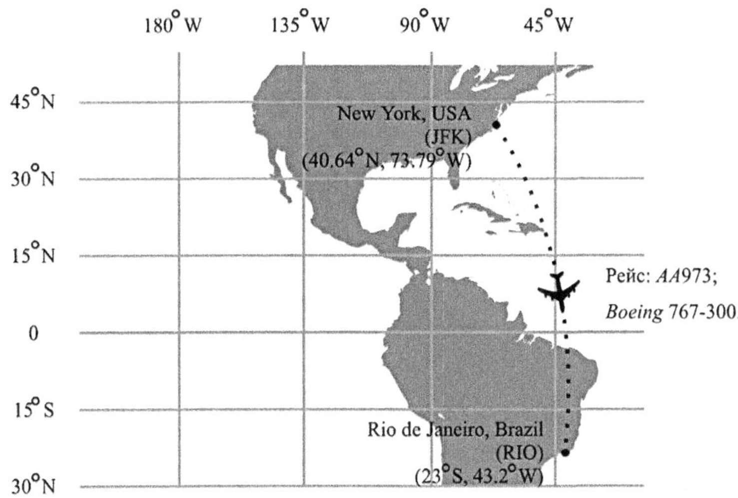

Fig. 3.6 Flight path of the American Airlines AA−973 route

by its amplitude and frequency values is two times greater than the same values of a strong magnetic storm. It has a frequency of a range of tens of microhertz, which is several times greater than the frequency values of a typical magnetic storm.

It is obvious that GMPS parameters depend on the region of flight vehicle exploitation and its flight characteristics (Table 3.9). Therefore, modern technological limits in the amplitude and frequency ranges of GMPS of 0–70,000 nT in amplitude and 0–3 mHz in frequency have been found. These values are at least three times greater than typical (traditional) geomagnetic variations.

It is useful to consider the metallic (duralumin) aircraft fuselage as a conductor (of cylindrical form), which is moving in an alternating and anisotropic field. According to the Maxwell equations, onboard there is an electromagnetic field of an appropriate frequency and amplitude, which is induced by the GMPS effect.

The range of the electromagnetic waves has a lowest level of 3 Hz (reaching 30 Hz at the highest level) as defined by the International Telecommunication Union (ITU). This range is known as an *extremely low frequency* (ELF). Nevertheless, it is suggested to be extended by a range of 0–3 Hz, which is defined as a *sub–extremely low frequency* (SELF).

Today, the problem of neutralization of the GMPS influence on biological and technical objects and systems onboard aircraft is still unsolved and not much studied. This fact poses new and complicated scientific and technical problems.

Table 3.8 Experimental data

No.	Latitude	Longitude	Altitude m/ft	$\boldsymbol{B}_X$, nT	$\boldsymbol{B}_Y$, nT	$\boldsymbol{B}_Z$, nT	$\boldsymbol{B}$, nT
1.	40.63°N	73.77°W	2/6	23,781	−4665	47,771	53,566
2.	40.07°N	73.22°W	01981/6500	23,955	−4776	47,181	53,129
3.	39.85°N	73.02°W	02682/8800	24,023	−4815	46,952	52,960
4.	39.58°N	72.76°W	3840/12,600	24,102	−4866	46,658	52,740
5.	39.22°N	72.39°W	4633/15,200	24,211	−4940	46,275	52,459
6.	38.87°N	72.01°W	6400/21,000	24,302	−5011	42,871	52,152
7.	38.49°N	71.71°W	7254/23,800	24,418	−5069	45,480	51,869
8.	38.02°N	71.24°W	9754/32,000	24,534	−5154	44,934	51,454
9.	37.71°N	70.96°W	10,698/35,100	24,619	−5206	44,594	51,204
10.	37.34°N	70.54°W	10,942/35,900	24,723	−5286	44,182	50,904
11.	36.80°N	70.10°W	11,033/36,200	24,885	−5371	43,625	50,510
12.	36.40°N	69.71°W	11,033/36,200	24,998	−5445	43,191	50,200
13.	35.86°N	69.18°W	11,033/36,200	25,147	−5543	42,598	49,776
14.	35.06°N	68.50°W	11,033/36,200	25,365	−5667	41,734	49,165
15.	*34.37°N*	*67.90°W*	*11,033/36,200*	*25,544*	*−5773*	*40,973*	*48,627*
16.	33.97°N	67.57°W	11,033/36,200	25,645	−5831	40,532	48,317
17.	32.61°N	66.52°W	11,033/36,200	25,978	−6011	39,032	47,270
18.	31.89°N	65.95°W	11,033/36,200	26,142	−6106	38,220	46,706
19.	30.95°N	65.13°W	11,033/36,200	26,342	−6239	37,124	45,945
20.	29.83° N	64.43°W	11,033/36,200	26,582	−6354	35,888	45,110
21.	28.60°N	63.53°W	11,033/36,200	26,821	−6497	34,471	44,157
22.	27.97°N	63.05°W	11,033/36,200	26,936	−6572	33,733	43,665
23.	26.78°N	62.30°W	11,033/36,200	27,150	−6693	32,379	42,783
24.	26.36°N	61.99°W	11,033/36,200	27,221	−6742	31,883	42,461
25.	25.95°N	61.71°W	11,033/36,200	27,289	−6787	31,405	42,154
26.	25.15°N	61.20°W	11,033/36,200	27,418	−6870	30,480	41,569
27.	24.47°N	60.78°W	11,033/36,200	27,524	−6940	29,694	41,079
28.	23.62°N	60.19°W	11,033/36,200	27,649	−7037	28,684	40,457
29.	22.97°N	59.82°W	11,033/36,200	27,741	−7102	27,937	40,006
30.	22.32°N	59.49°W	11,033/36,200	27,830	−7162	27,201	39,569
31.	21.71°N	59.14°W	11,033/36,000	27,909	−7225	26,494	39,154
32.	20.71°N	58.50°W	11,033/36,000	28,029	−7339	25,303	38,467
33.	19.86°N	58.04°W	11,033/36,000	28,123	−7426	24,316	37,912
34.	18.99°N	57.53°W	11,033/36,000	28,211	−7523	23,285	37,345
35.	18.22°N	57.18°W	11,033/36,000	28,279	−7597	22,408	36,872
36.	17.26°N	56.72°W	11,033/36,000	28,351	−7693	21,300	36,286
37.	14.49°N	55.29°W	11,033/36,000	28,473	−7992	18,023	34,633
38.	13.55°N	54.85°W	11,033/36,000	28,480	−8089	16,917	34,099
39.	12.55°N	54.30°W	11,033/36,000	28,467	−8201	15,699	33,527
40.	11.58°N	53.93°W	11,033/36,000	28,426	−8287	14,585	33,007
41.	10.41°N	53.45°W	11,033/36,000	28,346	−8393	13,222	32,384
42.	09.42°N	52.98°W	11,033/36,000	28,250	−8487	12,035	31,858

(continued)

Table 3.8 (continued)

No.	Latitude	Longitude	Altitude m/ft	B_X, nT	B_Y, nT	B_Z, nT	B, nT
43.	08.64°N	52.70°W	11,033/36,000	28,152	−8547	11,143	31,460
44.	07.31°N	52.10°W	11,033/36,000	27,946	−8657	9560	30,779
45.	06.44°N	51.67°W	11,033/36,000	27,782	−8729	8507	30,338
46.	05.43°N	51.31°W	11,033/36,000	27,559	−8788	7359	29,849
47.	04.14°N	50.83°W	11,033/36,000	27,229	−8855	5892	29,223
48.	02.89°N	50.37°W	11,033/36,000	26,861	−8906	4488	28,652
49.	01.75°N	49.91°W	11,033/36,000	26,483	−8945	3202	28,136
50.	00.65°N	49.56°W	11,033/36,000	26,085	−8961	2031	27,656
51.	02.14°S	48.59°W	11,033/36,000	24,935	−8956	−879	26,509
52.	04.55°S	47.91°W	11,033/36,000	23,819	−8874	−3158	25,614
53.	05.93°S	47.43°W	11,033/36,000	23,134	−8809	−4437	25,149
54.	07.81°S	46.90°W	11,033/36,000	22,178	−8682	−6019	24,565
55.	09.55°S	45.33°W	11,033/36,000	21,165	−8607	−7916	24,179
56.	11.49°S	45.84°W	11,033/36,000	20,258	−8352	−8793	23,611
57.	14.32°S	45.14°W	11,033/36,000	18,806	−8031	−10,569	23,019
58.	19.30°S	43.98°W	9754/32,000	16,387	−7387	−13,094	22,238
59.	20.09°S	43.89°W	8534/28,000	16,036	−7279	−13,408	22,134
60.	20.46°S	43.76°W	7010/23,000	15,868	−7236	−13,590	22,110
61.	20.91°S	43.71°W	5974/19,600	15,673	−7175	−13,762	22,057
62.	21.48°S	43.58°W	4572/15,000	15,421	−7101	−13,998	22,004
63.	21.97°S	43.54°W	3078/10,100	15,215	−7035	−14,173	21,957
64.	22.90°S	43.21°W	396/1300	14,800	−6920	−14,578	21,896

3.6 Algorithms of Geomagnetic Field Calculation

There are several methods to calculate undisturbed geomagnetic field parameters. Figure 3.8 presents an algorithm used for calculation of undisturbed geomagnetic field parameters for a single geographical location with given spatiotemporal coordinates. This algorithm has been realized in the numerical computational package *Scilab* (Alekseev et al. 2008; Vashenko 2008; Samarsky 1959). Here the geomagnetic field induction of potential intraterrestrial sources is calculated on the basis of given spherical coordinates r, θ, and λ.

The $\boldsymbol{X}'$, $\boldsymbol{Y}'$, and $\boldsymbol{Z}'$ components of the geomagnetic field induction vector are defined by Eq. 3.54, which includes the operations of the appropriate partial derivatives calculation:

$$\boldsymbol{X}' = \frac{1}{r}\frac{\partial U}{\partial \theta}; \quad \boldsymbol{Y}' = -\frac{1}{r \cdot \sin\theta}\frac{\partial U}{\partial \lambda}; \quad \boldsymbol{Z}' = \frac{\partial U}{\partial r}, \tag{3.54}$$

A great problem in the calculation algorithm concerns an effective way of calculation of the partial derivative in the given location. The most obvious solution to this is to apply numerical differentiation of the polynomial. According to this

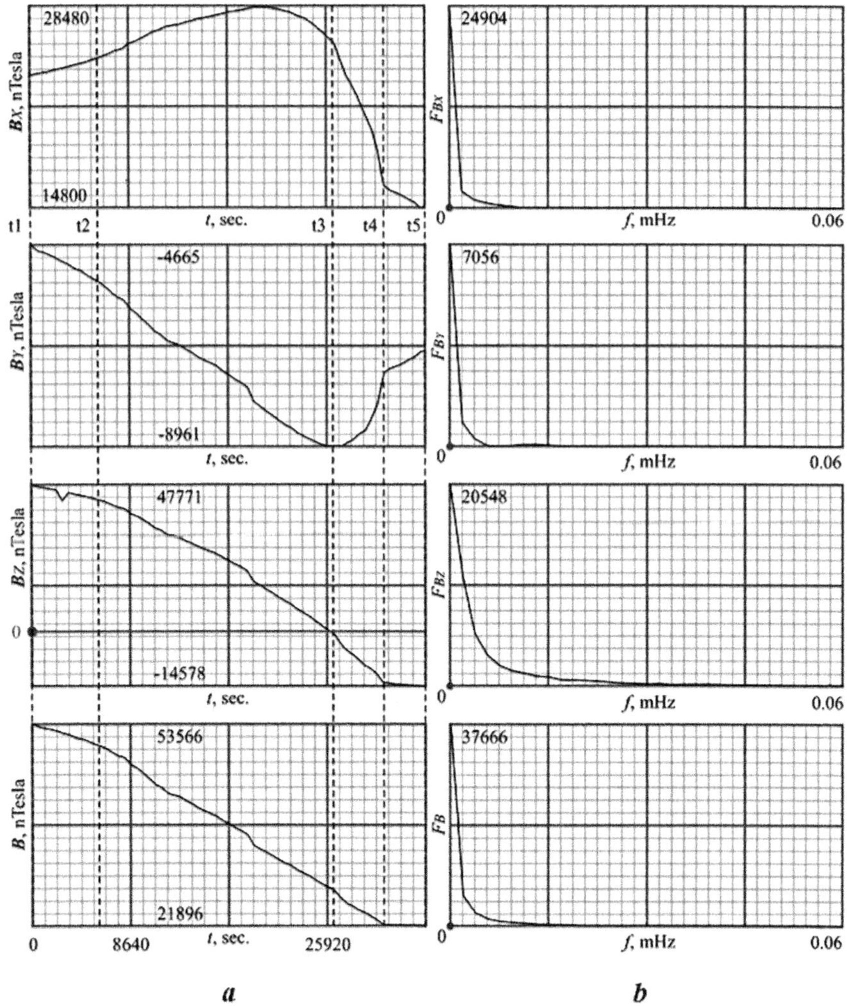

Fig. 3.7 Experimental data analysis results: amplitude (***a***) and frequency (***b***) ranges

Table 3.9 Flight characteristics of some aircraft

Aircraft	Ceiling, m	Speed, km/h	
		Cruise	Maximal
IL–96	12,000	870	910
Boeing 767–300	12,800	870	910
A 350–800	13,100	903	945
F–15	20,000	917	2650
Mig–31	20,600	2500	3000
X–43A	30,000	–	11,230

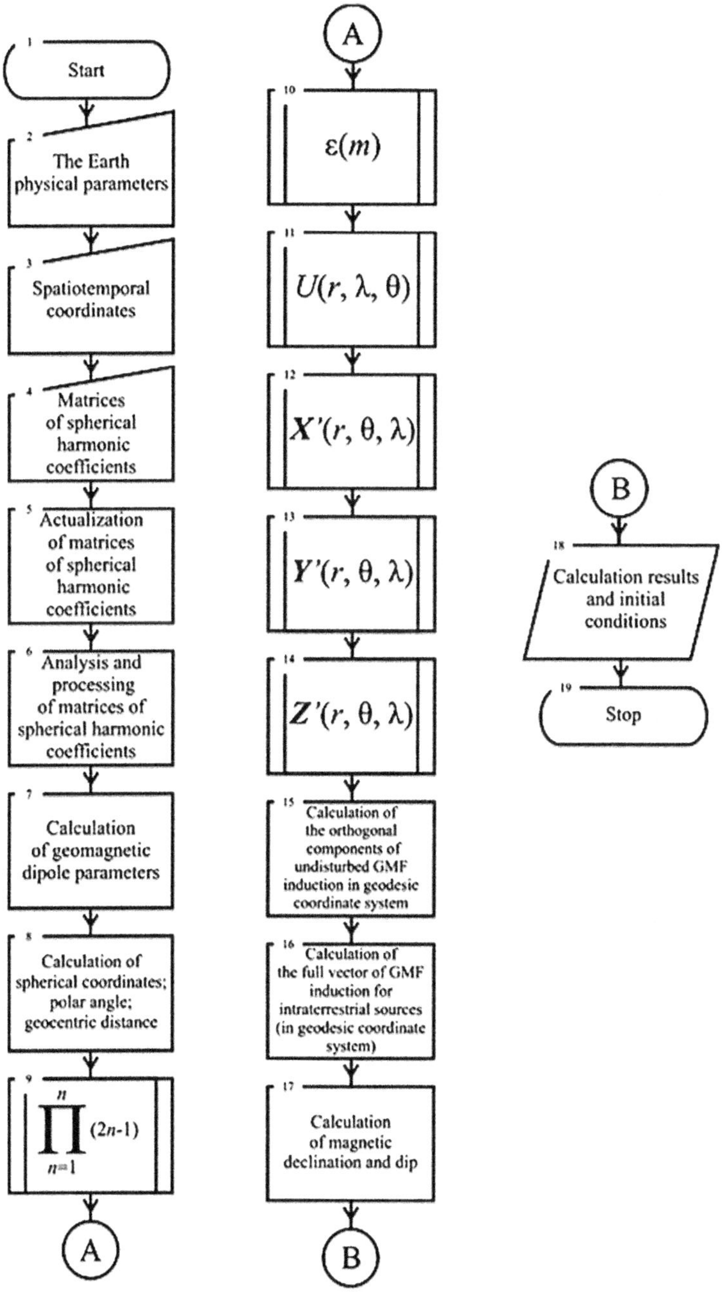

Fig. 3.8 Algorithm to calculate undisturbed geomagnetic field parameters for given spatiotemporal geographical coordinates according to (Standard GOST 19.003–80 1985a)

method, any function can be tabularly interpolated by the polynomial, which is used as an approximate value of the required derivative tabular function.

Newton polynomials are widely known for forward interpolation (Eq. 3.55):

$$y(x) = y_0 + q\Delta y_0 + \frac{q(q-1)}{2!}\Delta^2 y_0 + \cdots + \frac{q(q-1)\ldots(q-n-1)}{n!}\Delta^n y_0, \tag{3.55}$$

where $q = (x - x_0)/h$; $\Delta^k y_0$ are finite differences; and h is a step of argument and backward interpolation:

$$y(x) = y_n + q\Delta y_{n-1} + \frac{q(q+1)}{2!}\Delta^2 y_{n-2} + \cdots + \frac{q(q+1)\ldots(q+n-1)}{n!}\Delta^n y_0, \tag{3.56}$$

where $q = (x - x_n)/h$.

Newton polynomials can help to calculate partial derivative in case of polynomial interpolation for no more than 3–4 nodes.

The great number of nodes and the degree of the polynomial n justifies the oscillatory character of the polynomial. For example, let us use Eqs. 3.57 and 3.58:

$$\frac{\partial y}{\partial x} = \frac{\partial y}{\partial q}\cdot\frac{\partial q}{\partial x} = \frac{1}{h}\cdot\frac{\partial y}{\partial q}, \tag{3.57}$$

$$y'(x) \approx \frac{\Delta y_0}{h} + \frac{2q-1}{2h}\Delta^2 y_0 + \frac{3q^2-6q+2}{6h}\Delta^3 y_0 + \frac{2q^3-9q^2+11q-3}{12h}\Delta^4 y_0, \tag{3.58}$$

to find the first–order derivative in the point $x = x_0$ with Eq. 3.59:

$$y'(x_0) \approx \frac{1}{h}\left(\Delta y_0 - \frac{\Delta^2 y_0}{2} + \frac{\Delta^3 y_0}{3} - \frac{\Delta^4 y_0}{4}\right). \tag{3.59}$$

$$y''(x_0) \approx \frac{\partial(y')}{\partial x} = \frac{\partial(y')}{\partial q}\cdot\frac{\partial q}{\partial x} = \frac{1}{h}\cdot\frac{\partial(y')}{\partial q}, \tag{3.60}$$

The second–order derivative is defined by Eq. 3.61:

$$y''(x) \approx \frac{1}{h^2}\left(\Delta^2 y_0 - (q-1)\Delta^3 y_0 + \frac{6q-18q+11}{12}\Delta^4 y_0 + \cdots\right), \tag{3.61}$$

The second–order derivative of the function with tabular definition in the point $x = x_0$ is defined by Eq. 3.62:

$$y''(x_0) \approx \frac{1}{h^2}\left(\Delta^2 y_0 + \Delta^3 y_0 + \frac{11}{12}\Delta^4 y_0 - \frac{5}{6}\Delta^5 y_0\right). \tag{3.62}$$

An any–order derivative via the second Newton interpolation of the polynomial or Stirling's approximation is calculated in similar way. It should be mentioned that some typical truncation and rounding errors are possible in numerical derivative calculation. In practice, during the estimation of a truncation error, it is supposed that $f(x)$ does not have oscillating components for a period less than h. With this condition, a value of differences describes a quality of $f(x)$ function approximation by interpolation polynomial of the appropriate degree. When the differences of the m order are less than their rounding error, it is supposed that these differences are almost constant and a truncation error is less than the unit of the least significant digit of y_i/h values.

With the calculation step increase a truncation error decreases according to Eq. 3.63:

$$Q^1 = \frac{h}{2}f''(x). \tag{3.63}$$

The rounding error is inversely proportional to h in the first derivative calculation, inversely proportional to h^2 in the second derivative calculation, etc. During the increase of the calculation step h a rounding error increases. There are some rules from the theory of errors for the estimation, which can be defined by Eq. 3.64:

$$Q_2 = \frac{2\varepsilon \cdot f(x)}{h}, \tag{3.64}$$

where ε is an error of the function values definition.

The general error of derivative calculation can be considered in Eq. 3.65 as a sum of truncation and rounding errors:

$$Q = \frac{2\varepsilon \cdot f(x)}{h} + \frac{h}{2}f''(x). \tag{3.65}$$

With the interpolation order increase the truncation error decreases and the rounding error is increasing. So, Eq. 3.66 defines an optimal calculation step with minimal total error:

$$h_0 \approx \sqrt{4\left|\frac{\varepsilon \cdot f(x)}{f''(x)}\right|}. \tag{3.66}$$

According to this, Fig. 3.9 could be implemented with an algorithm that provides calculation of first– and second–order partial derivatives.

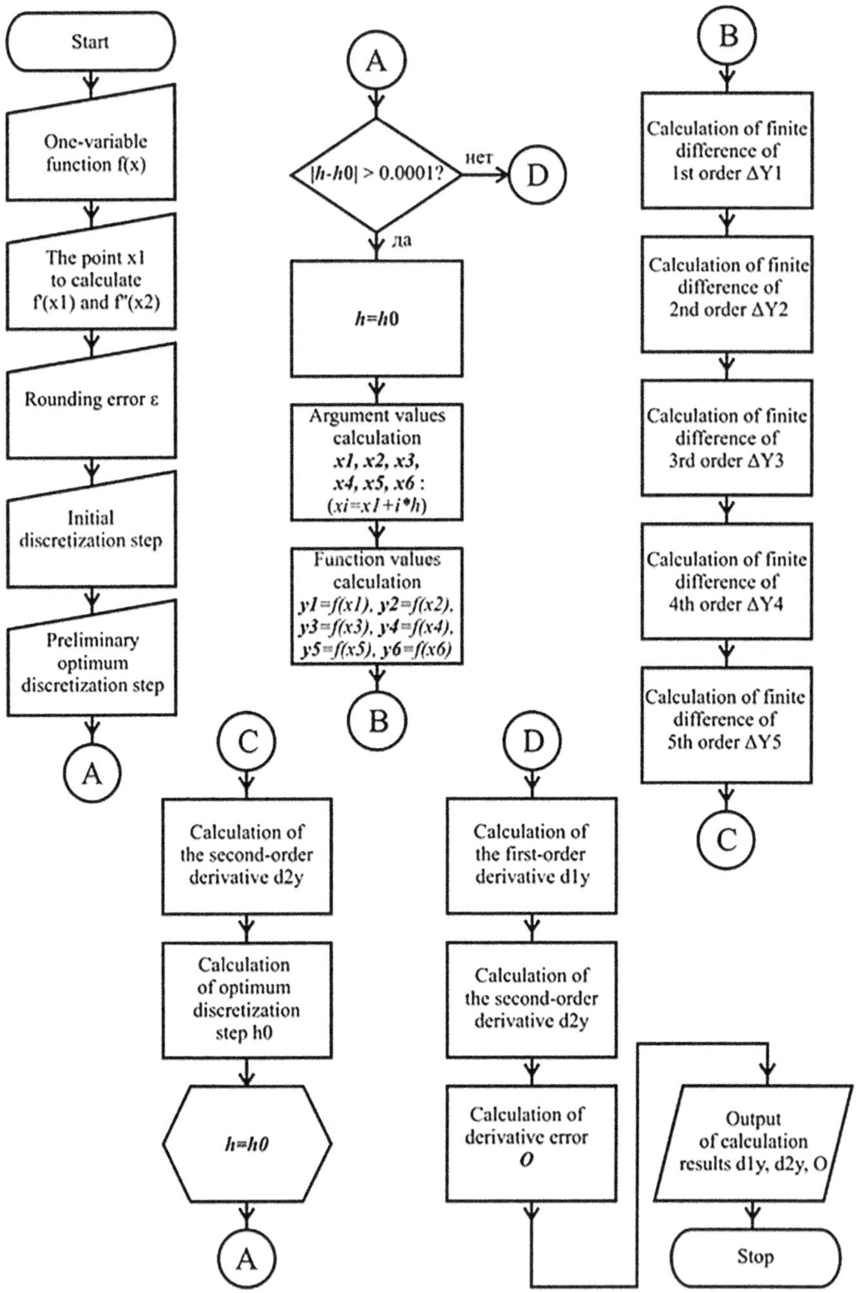

Fig. 3.9 Tabular representation of the algorithm used for calculation of the first– and second–order partial derivatives of a function

In fact, this algorithm provides a calculation of $\boldsymbol{X}'$, $\boldsymbol{Y}'$, and $\boldsymbol{Z}'$, geomagnetic dipole parameters and angle elements of the geomagnetic field that can be implemented on various software development platforms.

However, the components $\boldsymbol{Y}'$ and $\boldsymbol{Z}'$ of the geomagnetic field induction vector are defined by Eqs. 3.67 and 3.68:

$$\begin{aligned}\boldsymbol{Y}' &= -\frac{1/r}{\sin\theta}\frac{\partial U}{\partial\lambda}\\ &= \sum_{n=1}^{N}\sum_{m=0}^{n} m\left(g_n^m \sin(m\lambda) - h_n^m \cos(m\lambda)\right)\left(\frac{R_E}{r}\right)^{n+2}\frac{P_n^m \cos\theta}{\sin\theta};\end{aligned} \tag{3.67}$$

$$\begin{aligned}\boldsymbol{Z}' &= \frac{\partial U}{\partial r}\\ &= -\sum_{n=1}^{N}\sum_{m=0}^{n}(n+1)\left(g_n^m \cos(m\lambda) + h_n^m \sin(m\lambda)\right)\left(\frac{R_E}{r}\right)^{n+2} P_n^m \cos\theta;\end{aligned} \tag{3.68}$$

The introduction of Eqs. 3.67 and 3.68 does not require the use of a method of interpolation of polynomial numerical differentiation.

3.7 Methods of Analysis of Geomagnetic Field Parameters

Another important aspect of the programming solution is concerned with digital processing of data on space weather, the geomagnetic field, and its variation parameters. The approach is based on classic mechanisms of digital signal processing and supposes linear, nonlinear, and adaptive data filtering and analysis. Figures 3.10 and 3.11 represent these methods.

It is necessary to mention that calculation and exclusion of the constant component in step 3 is performed according to Eq. 3.69:

$$x_k^* = x_k - \frac{1}{N}\sum_{k=0}^{N-1} x_k, \tag{3.69}$$

where x_k^* is the $k-$th element of the array with a constant component excluded from the source array x_k; and N is the length of the array.

Next the standard deviation and variance of the information signal are calculated according to Eq. 3.70:

$$\sigma^2 = \frac{1}{N-1}\sum_{k=0}^{N-1}\left(x_k^*\right)^2, \tag{3.70}$$

where σ is a standard deviation and σ^2 is a variance of the information signal.

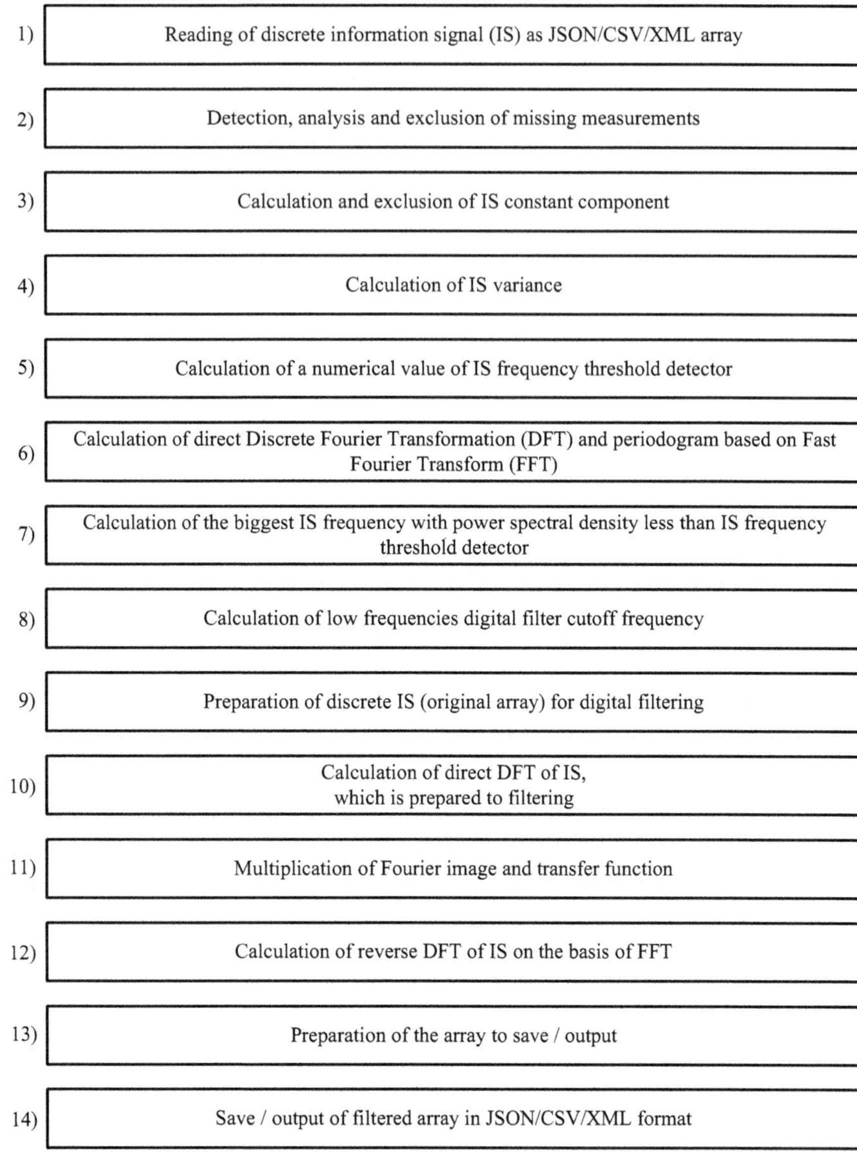

Fig. 3.10 Main methodic of analysis of geomagnetic field parameters

In step 5 (Figs. 3.10 and 3.11) a numerical value of the information signal frequency threshold detector l is calculated according to Eq. 3.71:

$$l = \frac{\sigma^2 X_q}{N}, \quad \text{where } X_q = -\ln(q). \tag{3.71}$$

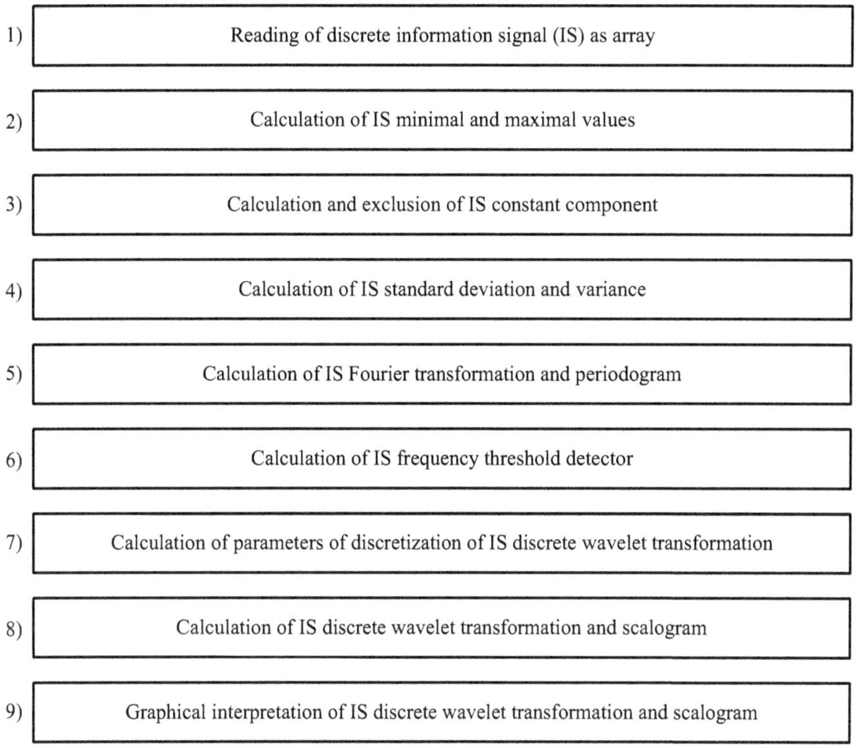

Fig. 3.11 Methodic of information signal time–frequency analysis

This value indicates that the harmonic exists in the information signal not as a component of white noise. The probability of such a statement is $1 - q$, where $q << 1$ (for example, in this case, $q = 0.03$).

Fourier transformation of the discrete function with an excluded constant component is used in step 6 (Figs. 3.10 and 3.11) to calculate an information signal periodogram with Eq. 3.72. In case of an odd N the upper limit of calculation is $(N + 1)/2$:

$$D_j = \frac{1}{N^2}\left[\left(\mathrm{Re}X_j\right)^2 + \left(\mathrm{Im}X_j\right)^2\right], \quad j = 0, 1, \ldots, N/2, \tag{3.72}$$

where D_j is the j–th element of the periodogram; and Re X_j and Im X_j are real and imaginary parts of the Fourier transformation, which are calculated as follows:

$$X_j =^j \sum_{k=0}^{N-1} x_k^* e^{-i\frac{2\pi}{N}kj}, \quad j = 0, 1, \ldots, N-1,$$

So, if Δt is a discretization step of the source signal, then the periodogram counts refer to the frequencies:

$$v_j = \Delta v \cdot j, \quad j = 0, 1, \ldots, N/2, \quad \text{where } \Delta v = \frac{1}{N\Delta t}.$$

Next, in step 7 (Fig. 4.7) the discretization parameters for discrete Fourier transformation are calculated. So, for example, the rounding up value of the maximal scale level $a_{\max}$ is calculated as Eq. 3.73:

$$a_{\max} = \log_2\left(\frac{N}{2}\right). \tag{3.73}$$

Next (Fig. 3.11) a discrete wavelet transformation of the information signal is calculated according to Eq. 3.74:

$$W(k, a_i, b_j) = \sum_{-N/2}^{N/2}\left[\frac{1}{\sqrt{a_i}} \cdot x_k^* \cdot \Psi\left(\frac{k - b_j}{a_i}\right)\right], \tag{3.74}$$
$$\text{where } a_i = a_0^m; a_0 = 2; \quad m = 0, 1, \ldots, a_{\max};$$

as well as its scalogram according to Eq. 3.75:

$$S(a_i, b_j) = \left|W(a_i, b_j)^2\right|; \tag{3.75}$$

where Ψ is a parent wavelet; a is a scale parameter; and b is a shift parameter ($b = -N/2, \ldots, N/2$).

Steps 9 and 13 (Fig. 3.10) are concerned with preparations of information signal to digital filtering and a source array to save/output. In both cases the preparation is to multiply each element of the array by value $(-1)^i$, where i is an array element index. This operation provides a harmonics count from the origin of the coordinate system.

3.8 Conclusions

- The geomagnetic field is a complex structured natural entity with ambiguous field characteristics, which is distributed in the Earth (and in near–Earth space) and interacts with both astronomical objects and objects/processes on the Earth's surface, in the subsoil, and in near–Earth space.
- A number of additional geomagnetic activity indices are suggested:
  - M–index and δM–index are respectively the absolute and relative deviation of the geomagnetic field at a particular moment of time at a defined geographical point from the analytically calculated value of geomagnetic field intensity at the same point.
  - δM–index and δM_Σ–index are respectively the absolute and relative deviation of the geomagnetic field at a particular moment at a particular location, which is defined by its spatiotemporal coordinates, from the value of

geomagnetic field induction as the geomagnetic field of intraterrestrial sources and geomagnetic anomalies in the region superposition at the same point.

- Common classification features of geomagnetic variations have been defined, and a classification scheme formulated to solve the problem of identification of geomagnetic variations.
- The main effects of the influence of geomagnetic variations on objects, systems, and processes of natural (especially human) and technical origin have been defined. A number of these influences' mechanisms have been explained.
- The relevance and the necessity of research for modernization of existing scientific methods and for creation of new ones concerning analytical calculation of the geomagnetic field and its variations has been highlighted, and the development and implementation of devices, tools, and systems with advanced characteristics for ecological, geophysical, and analytical estimation of the geomagnetic field and its variations are suggested.
- The main scientific and technical problems have been formulated and prospects for their solutions have been defined.
- An algorithm and software called *GEOmagnetic_v*1.0, which formalizes the mathematical model of intraterrestrial sources of geomagnetic field induction vector parameter calculation (Certificate of official computer program registration No. 2013610905), has been developed and certain ways of instance optimization and implementation have been proposed.
- Scientific and technical problems in increasing *GEOmagnetic_v*1.0 software efficiency and optimizing computer calculations have been formulated and solved.
- There is a method for automated generation of an undisturbed geomagnetic field induction value array and an appropriate magnetic isolines map both for the planet and the nearest ($h \leq 100$ km) circumterrestrial space and its local zones. The method takes into account the topographic features of the Earth's landscape in the studied regions.
- An approach for operative automated detection of undisturbed geomagnetic field parameters in conditions of minimum hardware and computing facilities and limited energy resources has been proposed.
- It has been proved that, while a biological object is moving on the Earth's surface or in its vertical plane, the anisotropy of the undisturbed geomagnetic field causes a natural magnetic influence, which is comparable to strong magnetic storms.
- The term *geomagnetic pseudostorm* (GMPS) has been defined and scientifically substantiated. The main parameters of the effect have been defined and a method for their calculation has been suggested.
- Detailed analysis has demonstrated that the amplitude and frequency characteristics of GMPS are comparable to—and, in some cases, greater than—geomagnetic variations, which are induced by the rotation of the Earth around its own axis or the solar wind. This means that GMPS can cause the same influence on biological objects as a strong magnetic storm.

References

Afanasiev YV, Studentsov NV et al (1968) Means of measuring the magnetic field parameters. Leningrad, Energiya, 320 p (in Russian)

Alekseev ER, Chesnokov OV, Rudchenko EA (2008) Scilab. Solving mathematical and engineering problems. Binom, Moscow. (in Russian)

Allegre CJ, Poirier JP, Humler E, Hofmann AW (1995) The chemical composition of the earth. Earth Planet Sci Lett 134(3–4):515–526

Belousov VV (1976) Geotektonikas. Moscow, MSU, 334 p (in Russian)

Belyavskaya NA (2004) Biological effects due to weak magnetic field on plants. Adv Space Res 34(7):1565–1574

Chizhevsky AL (1976) Terrestrial echo of solar storms. Moscow, Mysl, pp 5–15 (in Russian)

Dubrov AP (1978) The geomagnetic field and life. Springer, New York, ISBN:978-1-4757-1612-2 (Print) 978-1-4757-1610-8

Egeland A, Holter O, Lmholt A (1976) Space geophysics. Moscow, Mir, pp 3e1−33 (in Russian)

Fanselan G et al (1964) Die Darstellung des geomagnetischen Potentials zur Epoche 1945 durch eine Entwicklung nach Kugelfunktionen bis zur 15 Ordung. Pure Appl Geophys 57:5–30

Galland P, Pazur A (2005) Magnetoreception in plants. J Plant Res 118(6):371–389

Gavrilov VP (2005) Geotektonikas. Moscow, Oil and Gas, 368 p (in Russian)

Golicyn GS (1991) Modes of convection at different rotating geophysical and astro−physical objects. Izv USSR Acad Sci Atmos Ocean Phys 27(1):20−31 (in Russian)

Kampe de Ferrier J (1963) Functions of mathematical physics. Reference manual. Moscow, Physical and Mathematical Literature State Publishing House, pp 42–43 (in Russian)

Kush LA (1978) Bermuda Triangle: myths and realities. Moscow, Progress, 352 p (in Russian)

McDonough WF (1999) Earth's core. In: Marshall CP, Fairbridge RW (eds) Encyclopedia of geochemistry. Kluwer Academic Publishers, Dordrecht, pp 151–156

McDonough WF, Sun S−S (1995) The composition of the earth. Chem Geol 120:223–253

Merril RT, McElhinny MW, McFadden PL (1996) The magnetic field of the Earth. Academic, New York, 532 p

Panteleev VL (2000) Theory of earth figure. Moscow State University Publishing House, Moscow, pp 35–40 (in Russian)

Pudovkin MI, Raspopov OM, Kleymeova NG (1975) Perturbation of the electromagnetic field of the earth. P. 1. Leningrad University Publishing, Leningrad, pp 10–12 (in Russian)

Pushcharovskii YM (2001) Depths of the earth: the structure and tectonics of the mantle. Priroda 3:13–15 (in Russian)

Pushcharovskii YM, Pushcharovskii DY (1999) Geotektonika 1:3–14 (in Russian)

Russell CT (1993) Planetary magnetospheres (PDF). Rep Prog Phys 56(6):687–732. doi:10.1088/0034−4885/56/6/00

Samarsky A (1959) Introduction to numerical methods. Nauka, Moscow. (in Russian)

Schlegel K (2002) Füllekrug Martin: Weltweite Ortung von Blitzen: 50 Jahre Schumann−Resonanzen. Physik in unserer Zeit 33(6):256–261

Standard GOST 19.003−80 (1985a) Schemes of algorithms and programs. Graphical symbols. Moscow, Standards Publishing House (in Russian)

Standard GOST 25645.126−85 (1985b) Geomagnetic field. Model of field of intraterrestrial sources. Standards Publishing House, Moscow (in Russian)

Standard GOST 25645.127−85 (1985c) The Earth's magnetosphere. Model of magnetic field of magnetosphere currents. Standards Publishing House, Moscow (in Russian)

Vashenko GV (2008) Computational mathematics. Foundations of algebraic and trigonometric interpolation. Siberian State Technological University Publishing House, Krasnoyarsk, pp 4–6. (in Russian)

Vernadsky VI (2004) The biosphere and the noosphere. Iris Press, Moscow, pp 32–42 (in Russian)

Vorobev AV (2013a) A method of undisturbed geomagnetic field parameters detection in the field. Neftegazovoyedelo 1:71–80 (in Russian)

Vorobev AV (2013b) Modelling and research of geomagnetic pseudostorm effect. Geoinformatika 1:29–36 (in Russian)
Vorobev AV (2013c) GEOmagnetic_v1.0: certificate of computer program registration No. 2013610905
Vorobev AV. Approach to complex assess the effect of the geomagnetic pseudostorms: patent for invention No. 2526234
Vorobev AV, Garipova GT (2012) Prospects for modernization of control methods and parameters of electric power electricity mains powered systems. Neftegazovoyedelo 10(1):132–135 (in Russian)
Vorobev AV, Shakirova GR (2014a) Pseudostorm effect: computer modelling, calculation and experiment analyzes. In: Proceedings of the 14th SGEM Geoconference on informatics, geoinformatics and remote sensing, Albena, Bulgaria, 17–26 June 2014, vol 1, pp 745–751 (Scopus, doi:10.5593/sgem2014B21)
Vorobev AV, Shakirova GR (2014b) Geoinformation system of geomagnetic pseudostorm parameters registration and analysis. In: Proceedings of the 2nd international conference on Information Technologies for Intelligent Decision Support (ITIDS), vol 1, pp 160–165
Vorobev AV, Milovzorov GV, Milovzorov DG (2013) Methodics of geomagnetic pseudostorm parameters description. VestnikIzhGTU 1:103–107 (in Russian)
Vorobev AV, Shakirova GR, Khristodulo OI (2014a) To the problem of the geomagnetic pseudostorm effect modeling and analysis. In: Proceedings of the 2nd international conference on information technologies for intelligent decision, vol 3, pp 55–62
Vorobev AV, Biberacher M, Shakirova GR (2014b) Geoinformation system for geomagnetic pseudostorm parameters analysis and research. In: Proceedings of the 16th international workshop on computer science and information technologies, vol 1, Sheffield, England, 16–22 September 2014, pp 50–55
Vorobev AV et al (2014c) Analysis and research of special geomagnetic variations. Contemp Probl Sci Educ 2 (in Russian)
Weinstein SI (1983) Magnetic fields in space. Moscow, Nauka, pp 11–40 (in Russian)
Wiltschko W, Wiltschko A (2005) Magnetic orientation and magnetoreception in birds and other animals. J Comp Physiol A 191(8):675–693
Yanovsky BM (1978) Earth's magnetism. Leningrad University Publishing House, Leningrad, pp 87–88 (in Russian)

Chapter 4
The GEOMAGNET WebGIS Application

Modeling and visualization of the Earth's magnetic field is a challenging and complicated issue whose effective description and solution will contribute to better understanding of the principles and effects of geomagnetic field parameter distribution on the Earth's surface, in its subsoil, and in circumterrestrial space. All measured and collected data concerning the geomagnetic field are distributed in various sources and storages, so their integration into a single information means is of major importance. Nowadays, modern information technologies and capabilities provide a wide variety of tools for mathematical modeling and visualization using computer graphics that could play a key role also in the parameterization of the geomagnetic field and the modeling of its variations.

In this context, the authors propose and present a system that combines modern geoinformation techniques and Web technologies and provides the mechanisms to calculate, analyze, and visualize parameters of the geomagnetic field and its variations. The suggested system—developed on the basis of a set of conceptions, models, and methods—is called GEOMAGNET (https://www.geomagnet.ru) and provides effective integration of geomagnetic and space weather data into a single information means available to any specialist—a tool useful for operative monitoring, analytical control, modeling, and visualization.

4.1 Introduction

The parameters of space weather, the geomagnetic field, and its variations have a geospatial reference and can be classified as spatial data (Campbell 2003). Their distributed values strongly depend on geographical coordinates at any observation location. So, these data are described by the appropriate information technologies that provide a geospatial reference for attributive data and for their processing, analysis, and visualization (Amazon Web Services 2012; McLaughlin and Groot 2000).

S. Kolios et al., *GIS and Environmental Monitoring*, Geotechnologies and the Environment 20, DOI 10.1007/978-3-319-53086-4_4

One of the most effective approaches is based on using GIS technologies with their rich functionalities to process, analyze, and visualize geospatial data (Fig. 4.1). Most of the available models to represent the geomagnetic field parameter set distribution and their two-dimensional visualization results have been obtained using the GIS toolbox in ArcGIS software (Mapes 2008).

With this approach, continental, regional, local, or other types and models of geomagnetic field parameters at various scales and resolutions can also be realized. So, for example, Fig. 4.2 shows an example resulting from continental visualization of the main field (the geomagnetic field of intraterrestrial sources). The model has greater resolution than the model illustrated in Fig. 4.1 and has attributes similar to a geographic map.

In order to visualize both geospatial data and geomagnetic field parameters, a three-dimensional (3D) representation of the corresponding values can be used. In this case, GIS is more effective as it provides much more information and is able to

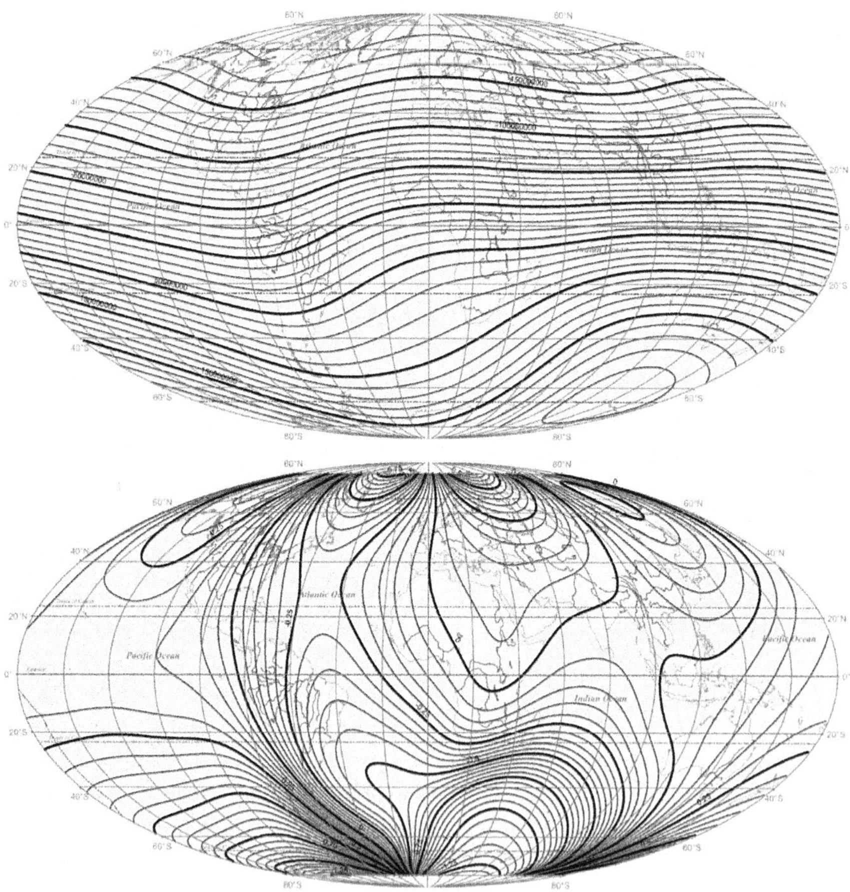

Fig. 4.1 Visualization of main field induction and magnetic declination distribution on the Earth's surface (nT · km)

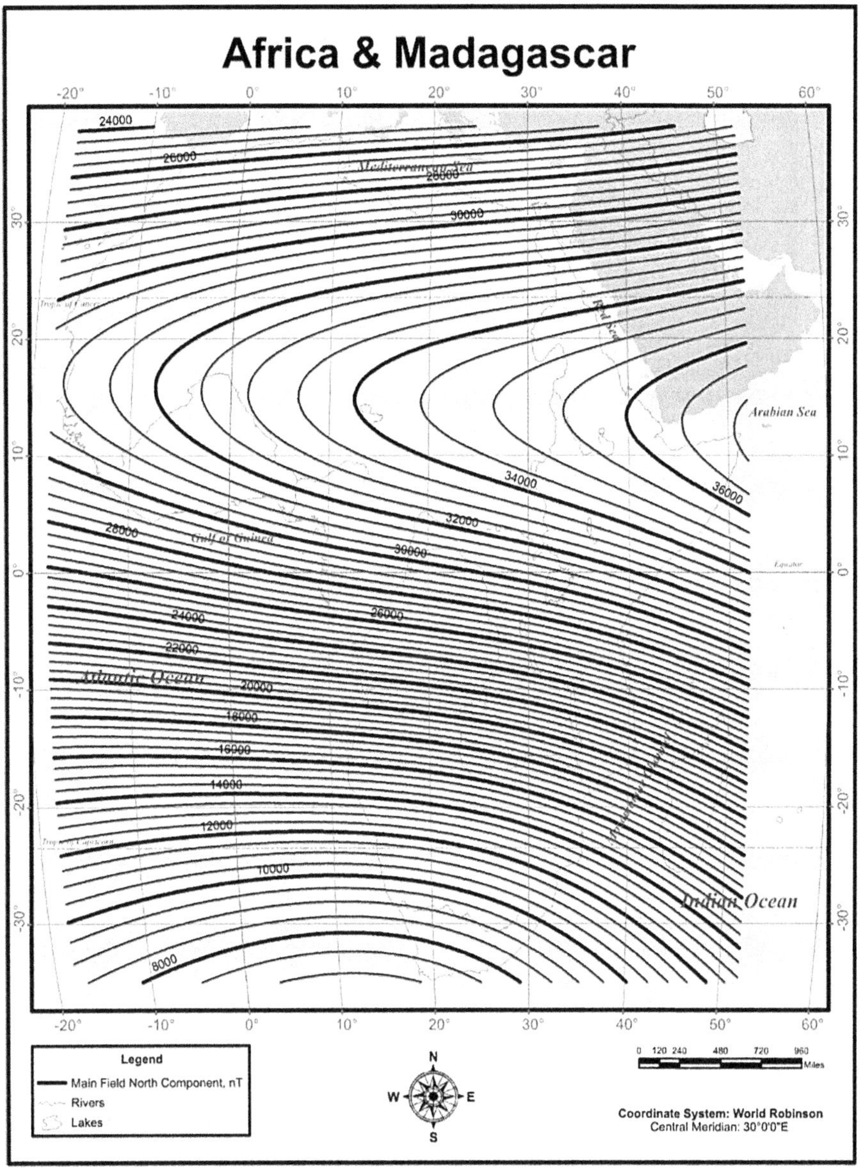

Fig. 4.2 Visualization of the main field B_X distribution in the Africa and Madagascar region (nT)

visualize the dynamic properties and multilevel scalability of calculated values (Samardak 2005). For example, Fig. 4.3 represents the unique results of a three-dimensional model of geomagnetic field parameter distribution on the Earth's surface. Similar models can be produced and synthesized for any other geomagnetic field parameters, depending on the corresponding geographic coordinate system.

However, no matter how powerful the toolset of GIS software is, the results are just a number of static images that can be used to study and/or analyze a particular

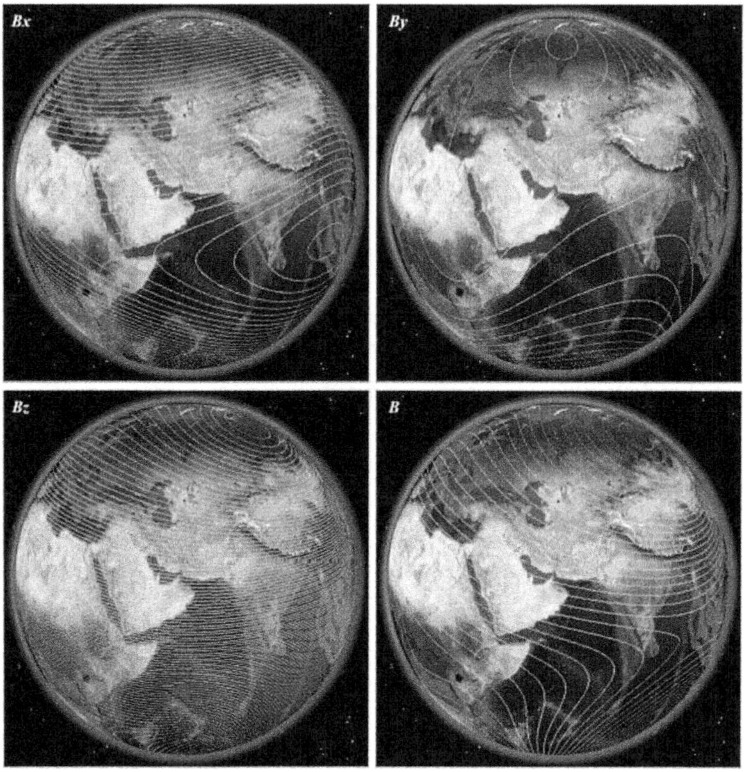

Fig. 4.3 Modeling and visualization of geomagnetic field parameters in 3D mode with a 1 μT step

parameter of the geomagnetic field at a given time. The preprocessing and production of these results require entering of the input data with a set of coordinates and parameters of the geomagnetic field in the appropriate geographic location. This procedure requires an expert in both using the complicated GIS software and exploring, interpreting, and analysis the geomagnetic field (Hall and Leahy 2008; Harvey and Han 2009; Longley et al. 2001; Peng and Tsou 2003; Yang et al. 2011; Zhurkin and Shaytura 2009). It is important that the end user can get any type of geomagnetic field analysis and visualization at any time and place just by a couple of mouse clicks, which is possible by using WebGIS technologies (Alesheikh et al. 2002; Carver 2001; Gupta and Sharma 2014; Haklay et al. 2008; Kim and Kim 2002).

4.2 Data Used

Nowadays, the geomagnetic field and its variation parameters are recorded on land, at sea, in the air, and in space. The issue here is that almost every type of geomagnetic data source has its own specific data format and presentation features.

So, to process them it is necessary to make some initial steps and convert them into a common format (Mandea and Korte 2011).

The majority of observations of the geomagnetic field are performed on land with a number of magnetic stations or observatories that belong to scientific organizations for parametric and astronomical observations of the Earth's magnetosphere. Modern magnetic station resolution is good enough and takes values up to 0.001 nT (Merrill et al. 1996).

Today, there are about 127 magnetic stations all around the world. All of them are under the jurisdiction of various scientific organizations, such as the Russian Academy of Sciences and the British Geological Survey. The majority of the stations are located in Europe (Figs. 4.4 and 4.5).

To determine the full vector intensity (induction) of the geomagnetic field, the magnetic observatories use magnetic theodolites and variational stations, which record geomagnetic field parameters in the automated mode. The recorded information is regularly sent to the international centers in Russia, the USA, Denmark, and Japan, where the information is recorded, analyzed, and made partially available to a broader audience with some delay.

A great example of geomagnetic field observations on land is the INTERMAGNET system—a global network of cooperating digital magnetic observatories (http://www.intermagnet.org/index-eng.php). The system has five international data centers, which collect data recorded by a network of magnetic stations, but in Russia, for example, there are only eight INTERMAGNET observatories and the volumes of the recorded data are not enough to describe the geomagnetic field parameters for the whole country.

The same applies to the recording of geomagnetic field parameters at sea and in the air. These observations are usually private ones, so they are not available to everyone. Data are collected for a limited territory (no more than 100 km) for some special use, such as geology or geophysics (for example, the magnetic station IMAGE in Greenland), and are suitable only for specific purposes.

Another way to record parameters of the geomagnetic field is more global and involves applying satellite observations of the magnetosphere. This is the case of the project called SWARM, which was initiated by the European Space Agency in

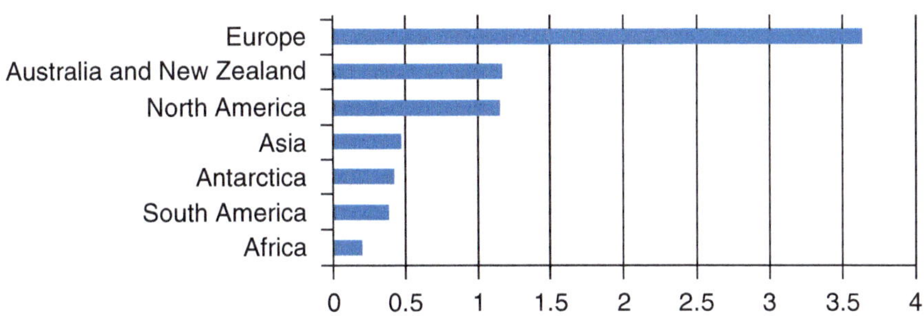

Fig. 4.4 Distribution of magnetic observatories on the Earth's surface (number of magnetic stations/million km^2)

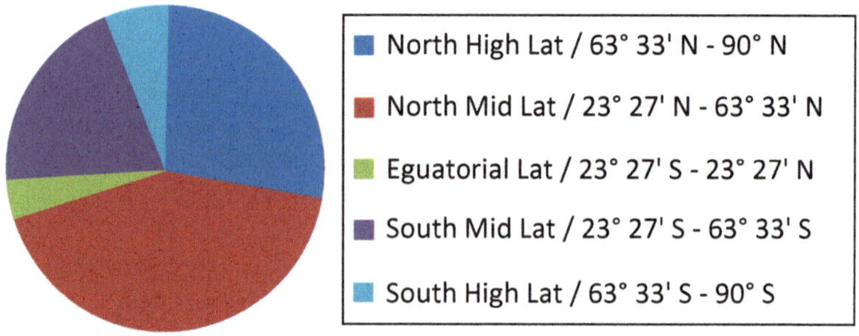

Fig. 4.5 Statistics on the distribution of magnetic stations by latitudinal geographic zones

2013 and involves three satellites, which record parameters of the Earth's magnetic field in real-time mode and send the collected data to specialized information centers. Another important space data source is the Geostationary Operational Environmental Satellite (GOES) system of the United States National Environmental Satellite, Data, and Information Service (NESDIS). This system records and monitors GMF parameters along with data on space weather.

Another class of maps, which is very important for scientists and specialists, concern the visualization of geomagnetic anomalies, which are described on the basis of observations and empirical estimation of the magnetic properties of rocks.

The necessary data to represent geomagnetic anomalies can be gathered by proton magnetometers, which are specific instruments depending on the measurement type: satellite, marine, airborne, or land.

The majority of geomagnetic data centers provide open access to information. However, these data are understandable only by specialists in geomagnetism or space weather, and not by other users. Moreover, it has to be mentioned that all data are distributed by various sources and it takes time to collect them and gather them together for analysis.

4.3 Conception of the GEOMAGNET WebGIS Application

In order to model and visualize the geomagnetic field parameters, the authors of this chapter (Vorobev and Shakirova 2015b) designed and developed the GEOMAGNET Web portal (https://www.geomagnet.ru), which provides complex calculation, analysis, and 2D/3D visualization of the geomagnetic field and its variation parameters. The geomagnetic field and its variation models, which are represented and described by GEOMAGNET, meet the requirements of specialists in various areas (geology, geophysics, medicine, meteorology, etc.). They

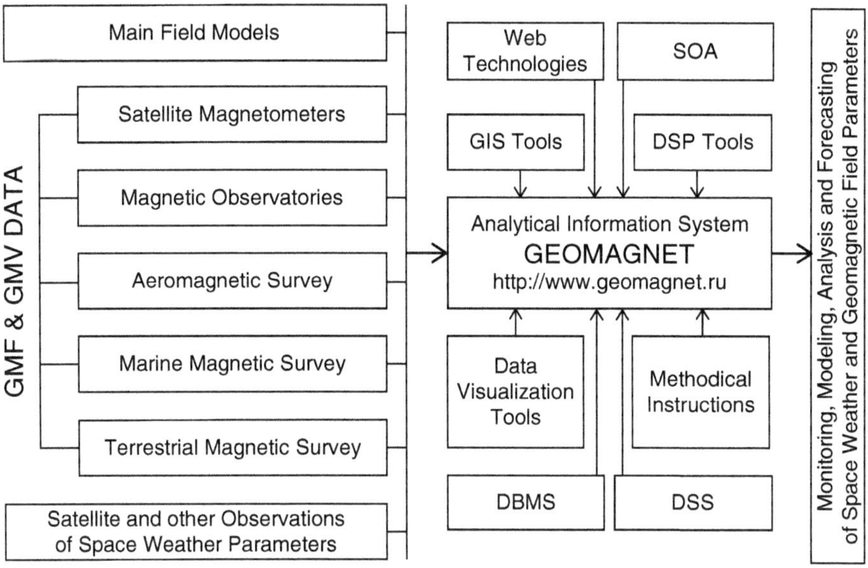

Fig. 4.6 Conception of the GEOMAGNET Web portal (https://www.geomagnet.ru)

effectively provide formatting and structuring of the data on the Earth's magnetosphere parameters and their further analysis (De Smith et al. 2007; Soloviev et al. 2013). With the Web portal, a user can get access to data, automatically analyze them, and further use the results in their applications and studies.

Figure 4.6 illustrates the conception of the system developed for monitoring, analyzing, and visualizing the parameters of space weather and the geomagnetic field and its variations. It is based on a paradigm of combining heterogeneous data sources into an integrated information space with a platform-independent mechanism for analytical control and interpretation.

Data from various heterogeneous sources (magnetic observatories, satellites, marine and terrestrial magnetic surveys) are integrated into the information system. The user is able to study, analyze, and visualize the data with a wide range of mechanisms: from GIS technologies to elements of service-oriented architecture (SOA) and digital signal processing (DSP) tools (Bogoutdinov et al. 2010; Flanagan 2011; ESRI Inc. 1999; Goodchild et al. 2007; Goodchild 2008; Green and Bossomaier 2002).

In general, the GEOMAGNET Web portal provides the main mechanisms for real-time monitoring, modeling, analytical control, and forecasting of space weather and the geomagnetic field and its variation parameters (the discretization step is 1 min). The Web portal provides all the available data and it is characterized by its informative capabilities, efficiency, and ergonomic criteria in monitoring and analytical study and presentation of space weather and the geomagnetic field and its variation parameters.

4.4 Characteristics and Services of the GEOMAGNET WebGIS Application

The main page of the GEOMAGNET Web portal is given in Fig. 4.7. The Web portal consists of three main services for modeling and analyzing space weather and the Earth's magnetic field parameters:

- *Geomagnetic Calculator* for calculating parameters of the normal geomagnetic field at a given point
- *Solar Activity* for analyzing 3-day space weather parameters
- *Magnetic Stations* for integrating data recorded by magnetic stations all around the world

All three services use modern WebGIS technologies for processing and visualizing geospatial data as well as a set of mathematical models and algorithms, described in Chap. 3.

The GEOMAGNET WebGIS application provides and is based on the following GIS mechanisms:

- Direct and reverse geocoding of spatial data (Tsvetkov 1998)
- Interpretation and visualization of spatial data
- Geolocation (identification of exact device location) (Svennerberg 2010)

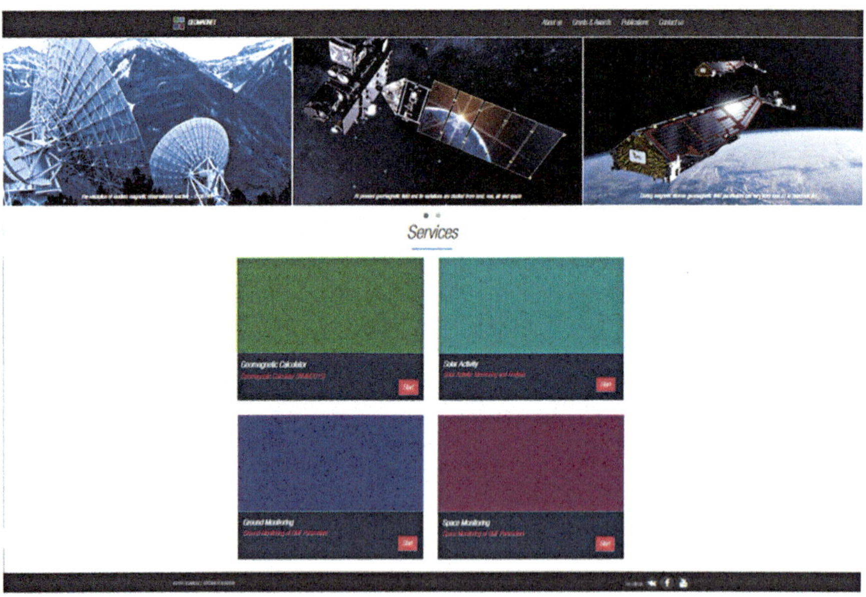

Fig. 4.7 Main page of the GEOMAGNET Web portal (https://www.geomagnet.ru)

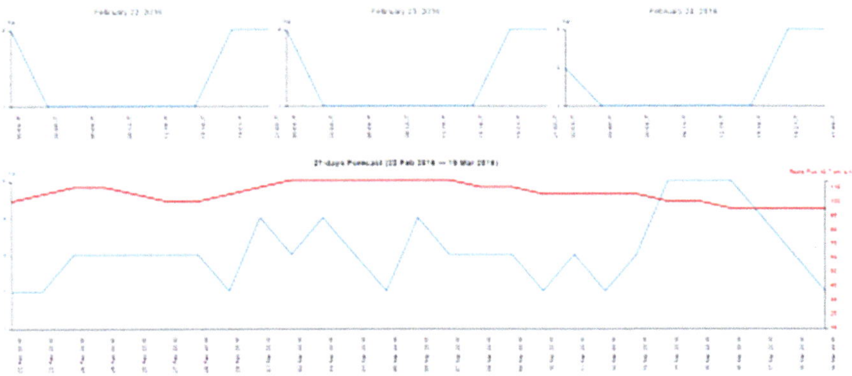

Fig. 4.8 Interpretation of geomagnetic variations forecast for 3 and 27 days (data provided by NOAA) by the GEOMAGNET Web portal

- Processing, visualization, translation, and analysis of available magnetic observatory data (with support from INTERMAGNET and the Geophysical Center of the Russian Academy of Sciences)
- Virtual globe based on WebGL technology (Gore 1999)

The GEOMAGNET Web portal also supports the following mechanisms of data visualization and interpretation:

- Multilayer visualization of spatial data with optional support for user-defined KML/KMZ layers
- Visualization of a two-dimensional array with dynamic time axis scalability on the basis of the Web interface D3.js Data-Driven Documents
- 2D/3D visualization of geospatial data with a scale-dependent detailization level, which is described by Eq. 4.1:

$$n = \log_2(\log_4 m) + 2, \tag{4.1}$$

  where m is the quantity of tiles; and n is the discrete scale level
- Forecasting (detailed for 3 days and daily averages for 27 days) of space weather and geomagnetic field parameters (Fig. 4.8)

4.4.1 The Geomagnetic Calculator (WMM2015) Service

The most common issue regarding the analysis of the Earth's magnetic field concerns calculation of the parameters of the normal geomagnetic field at a specific point with given geographical coordinates. This is calculated by applying some special methods known as geomagnetic calculators.

A geomagnetic (or magnetic) calculator is a special type of Web application, which provides an estimation of the magnetic field at a given location on Earth.

Here is a list of the most popular and widely used such services:

- NGDC NOAA (http://www.ngdc.noaa.gov/geomag-web)
- INTERMAGNET (http://www.intermagnet.org/data-donnee/data-eng.php)
- British Geological Survey (http://www.intermagnet.org/data-donnee/data-eng.php)
- US Geological Survey (http://geomag.usgs.gov)

All of them provide the simplest functionality. When a user enters a location, the application estimates the parameters of the geomagnetic field at the specific point. However, there is a common limitation in these applications: although they provide simple input fields for the geodetic coordinates of the point, the user has to know the exact parameters of the specific location, and this is impossible sometimes and thus precludes use of these tools.

Another issue is linked to the reliability of the calculation results. It is well known that the parameters of the geomagnetic field depend on three values: the latitude, longitude, and altitude of the point. All available application tools define the default altitude as being 0 km by default, so the results of the calculation are correct just for this elevation. A user has the potential to enter the exact altitude value using the input field, but most users (except professionals) do not use this, so this value is always referred to as 0.

To improve geomagnetic calculators, the authors suggest adding the following features:

- *Maps, e.g., geographical information technologies.* The geomagnetic field is estimated using geodetical coordinates for a given point on Earth. This class of data is referred to as geospatial and is traditionally processed by geoinformation systems and technologies. Using the map feature of the application, a user can simply pick the point of interest by mouse clicking.
- *Altitude calculation.* To increase the accuracy of calculation, it is necessary to provide an actual value of the altitude for a given point. The best solution is to get the value automatically on the basis of the geodetic latitude and longitude of the point. However, this feature should not exclude the possibility of inserting data manually in the case that one needs to observe how the magnetic field changes with elevation.
- *Visualization.* The most popular approach to visualize the parameters of the geomagnetic field is a set of contours. Each contour is a curve along which the parameter of the geomagnetic field has a constant value. A user chooses a parameter to be visualized and the system renders on the map a set of contours, which represent the distribution of the magnetic field on the Earth's surface.

- *Three-dimensional representation.* In this case, a GIS provides much more information than any other system or technology, and it is even more important due to the dynamic properties and multilevel scale ability.

Within the abovementioned context, and in order to provide the defined requirements, the authors (Vorobev and Shakirova 2015b; Vorobev and Shakirova 2015c) have suggested a special service: the *Geomagnetic Calculator (WMM2015)* (https://www.geomagnet.ru/WMM/index.html), which provides complex calculation of parameters of the geomagnetic field at a specific location with given spatiotemporal coordinates.

The Geomagnetic Calculator (WMM2015) is a Web-based application within the GEOMAGNET WebGIS application providing calculation of the parameters of the geomagnetic field and its secular variations according to a set of coordinates and dates (Fig. 4.9). A user at any level can calculate and analyze parameters of the geomagnetic field at his current location or at any other point on Earth (Vorobev and Shakirova 2015a; Vorobev et al. 2014).

The Geomagnetic Calculator (WMM2015) is a responsive application, which is transferable and works similarly on any device, i.e., phones, tablets, and desktops. This platform and device independency is realized on the basis of the special Bootstrap framework (http://getbootstrap.com) of programming libraries for HTML, CSS, and JavaScript (JQuery) (see also Sect. 1.13.2 in Chap. 1). The flexibility and performance of this application tool are also increased by supporting HTML5 and CSS3 standards.

The main functionality of the Geomagnetic Calculator (WMM2015) provides effective and reliable calculation and analysis of parameters of the normal (undisturbed) geomagnetic field by spatiotemporal coordinates with an error value of no more than 0.1% (Alekseev et al. 2008).

The interface of the Geomagnetic Calculator (WMM2015) service consists of two functional parts—panels (Fig. 4.9). The first (and the biggest) panel was

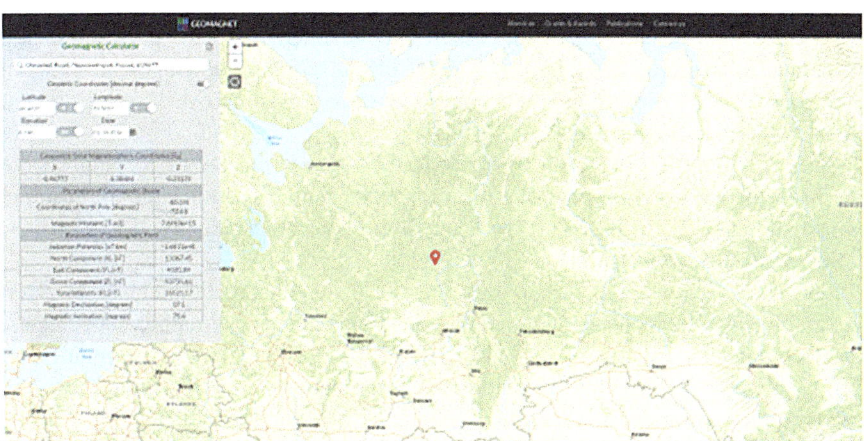

Fig. 4.9 Interface of the Geomagnetic Calculator (WMM2015)

developed for map rendering. All manipulations with geospatial data are visualized in this area (markers, zoom, geolocation, contours, etc.). The second panel is used to display information about both initial parameters (spatiotemporal coordinates) and calculated parameters (characteristics of the geomagnetic field), and it also presents the coordinates of the mouse pointer while moving over the map.

To calculate the parameters of the geomagnetic field, a user has to define the spatiotemporal coordinates at the point of interest on the Earth's surface. In Geomagnetic Calculator (WMM2015) the geodetic latitude and longitude can be defined in various ways.

The simplest way to define the current geographical position of the user is a geolocation function (Vorobev and Shakirova 2015a, b, c). This function automatically takes the IP address of the device that accesses the Internet. This functionality allows the user to get his location without searching for it on the map or filling in the appropriate input fields.

Another way to define a point is to pick it on the map. A user can move over the map (using a keyboard or mouse) and click on a specific point. All necessary spatiotemporal parameters of the point are calculated automatically and are immediately displayed on the screen.

Both aforementioned ways of location definition do not require the exact coordinates of the location as all necessary geospatial data are obtained automatically. A good, though more complicated, way to calculate parameters of the geomagnetic field and present a specific point on the map with high accuracy (up to a few meters) is by manually entering the coordinates into the input fields and thereafter the application displays the point on the map and the parameters of the normal geomagnetic field.

In addition, the Geomagnetic Calculator (WMM2015) provides a geocoding functionality to define a specific location. Geocoding is the transformation of an address of full form, i.e., "postal code, country, city, street, building number" or of short form, i.e., just one item or their combination, into the coordinate set "north latitude, east longitude". Thus, in order to provide the point to be calculated, the user can just enter in the "Search" field the address of the location he is interested in.

The Earth's surface point for which the geomagnetic field parameters are calculated is marked with a special marker on the visible map fragment. When a user picks a new point on the map to analyze it, any existing overlays are deleted (markers, info windows, etc.).

After defining the point for a search (either by coordinates or picking on the map), in the information panel there is a special field presenting the appropriate object address. This is based on the geocoding service (specifically via its realization—reverse geocoding).

As already mentioned, the parameters of the geomagnetic field at a given geographical point depend on its elevation. For defined geographic coordinates, the Geomagnetic Calculator (WMM2015) gets the altitude value automatically. The result is transformed either into the International System of Units or into the Imperial/US customary measurement system. In the API, the application includes a special service, called Elevation Service, which provides elevation data for all locations on the Earth's surface. It is a procedure with asynchronous interaction: after sending a request to the server, a user (or a Web page) does not wait for its response and continues performing all existing operations (or starts new ones) in background mode.

An important feature of the Geomagnetic Calculator (WMM2015) is data representation in one of two formats: *DD* (*decimal degrees*) and *DMS* (*degrees – minutes – seconds*). Depending on the chosen format, a user gets the appropriate input mask and the application supports the automatic transformation of coordinate systems via checking the corresponding radio button (next to the "Geodetic Coordinates" title on the information panel).

Parameters of the geomagnetic field are calculated according to a mathematical model represented in Chap. 3. Its programming realization is a set of functions to calculate parameters such as:

- North component of the geomagnetic field induction vector
- Vertical component of the geomagnetic field induction vector
- Magnetic declination and dip
- Scalar potential of the geomagnetic field induction vector

The results of the GEOMAGNETIC FIELD parameters complex calculation are presented in the International System of Units. This allows analysis of the data without any preliminary calculations.

All calculated results are available on-screen. Also a user can download a report (Fig. 4.10) by using the "Print" button in the top right corner of the information panel.

4.4.2 *The Solar Activity Service*

Geomagnetic variations are mainly caused by events of solar activity (space weather), whose parameters are measured in 24/7 mode by satellites and ground observatories. However, in this case, researchers have to deal with a lot of heterogeneous data that are hard to collect and use. Real-time mode monitoring of space weather is provided by another service on the GEOMAGNET Web portal called Solar Activity (https://www.geomagnet.ru/solar_activity.html).

The first question to investigate is what the characteristics of space weather are and why it is important to study this subject.

The first official definition of the term "space weather" was given in 1995 in the USA during development of the National Space Weather Program (NSWP). According to the definition, "space weather" is the set of changes on the Sun and in the solar wind, magnetosphere, and ionosphere that can influence onboard and ground technical systems' efficiency and reliability as well as harming human life and health. The most common recorded parameters of the solar flux are the velocity, temperature, and concentration of the solar wind.

GEOMAGNET
https://www.geomagnet.ru

mail@geomagnet.ru, geomagnet@list.ru

REPORT

Geodetic Coordinates

Latitude	Longitude	Elevation	Date
60.6027°N	52.5057°E	0.180 km	01-10-2016

Geocentric Solar Magnetospheric Coordinates [Radius of Earth]

X	Y	Z
-0.96777	0.38484	-0.23175

Parameters of Geomagnetic Dipole

Coordinates of North Pole	Magnetic Moment
-80.3°N, -72.6°E	7.6910e+15 T·m³

Parameters of Geomagnetic Field

X	Y	Z	F	D	I
13367.45 nT	4101.89 nT	53731.61 nT	55521.17 nT	17.1°	75.4°

Generated by GEOMAGNET at 01.10.2016, 20:57:36

Fig. 4.10 Report generated by the Geomagnetic Calculator (WMM2015)

In recent years, the new term "space climate" has been introduced, referring to the long-period variations of solar activity and space weather. Due to the dynamic development of space programs, an analytical description of space climate parameters has become increasingly important.

Actually, space weather parameters can be estimated by both ground-based and satellite-based observatories. The data resulting from the measurements are stored based on their heterogeneous sources—relational databases, textual files, image data, etc. However, the majority of the data are available via Web services provided, for example, by the NOAA Space Weather Prediction Center (www.swpc.noaa.gov/SWN).

The Web application Solar Activity: Monitoring and Analysis allows users to monitor solar weather in real time on the basis of satellite observations and carry out a comprehensive analysis of its parameters, to assess how their values change for different periods of time (from 1 h to several days). The service takes data via asynchronous requests to the Web server and renders results in real-time mode both in textual and in graphic forms. Data analysis is also available in amplitude and frequency mode by single or groups of parameters of space weather.

With the Solar Activity service a user gets the latest 3-day information on space weather parameters. In particular there are three main available data parameters:

- Proton density (cm^{-3})
- Temperature of the solar wind (K)
- Velocity of the solar wind (km/s)

Figure 4.11 represents the main window of the service. It is divided into two main parts: statistical and graphical.

In the statistical part the results of signal analysis in the time domain are illustrated. It contains information about values such as maximal, minimal, and average values, and variance and standard deviation, which are calculated for the proton density, temperature, and velocity of the solar wind. Also, the current values of these parameters are also displayed here.

Also, this part of the Solar Activity window estimates the characteristic of the magnetic state according to the Kp-index scale. According to this scale there are five possible magnetic states:

- Minor magnetic storm
- Moderate magnetic storm
- Strong magnetic storm
- Severe magnetic storm
- Extreme magnetic storm

These variables describe how current and future space weather conditions could influence various systems, especially human-based ones. Each one has a corresponding color: from green to deep red. The same color is used in the Solar

Fig. 4.11 Main Solar Activity window

Activity interface for indicating the current geomagnetic state. For example, the state "Below Geomagnetic Storm" with $Kp = 4$ is colored yellow (Fig. 4.12).

Another part of the Solar Activity window is a graphical one. It provides a set of charts that display the results of linear and nonlinear filtering of space weather data in the time and frequency domains (Figs. 4.13 and 4.14), as well as spectral and time–frequency analysis (Fig. 4.15).

The basic part of the data analysis in the Solar Activity service concerns the time domain (in UTC). The chart displays how the proton density, temperature, and velocity of the solar wind have changed over a 3-day period (Fig. 4.13).

The service includes a special feature where a user can choose any time period within the given 3-day domain to study the results of analysis in more details. Choosing a concrete time period via the chart brushing function, it allows selecting a region using the mouse or touch and zooming in on the appropriate area of the chart (Fig. 4.16).

Displaying "raw" data in the chart supposes that the information signal has some noise values in the series, thus impeding data analysis. As a solution to this, the Solar Activity service provides linear and nonlinear data filtration options (via the "Filtered" option in the top left corner of the service window) (Fig. 4.17). The optimal bandwidth values for the filtration are estimated automatically and displayed to the user on page loading with all default (unfiltered) charts (for each analyzed parameter of the solar wind).

At the same time, data filtering in the time and frequency domains also supposes adaptive filtration of the information signal. In this case, a user can manually enter the bandwidth value for the proton density, temperature, and velocity of the solar wind, and the service will provide the appropriate results (Fig. 4.18)

Another issue concerns possible missing values in the time series (for example, if the signal from a satellite was lost for a couple of minutes). To cope with this issue, the Solar Activity service also provides a possibility for analysis of the correlation of adjacent information signal parameters. Due to this possibility in the service, there are no missing chart areas, which could affect data analysis results.

A user can see the concrete values of analyzed parameters of the solar wind via special tooltips, which become visible on mouse movement over the chart (Fig. 4.19).

Spectral and time–frequency analysis in Solar Activity is implemented in two ways. The first one estimates the power spectral density via a periodogram. This approach is based on calculation of the data sequence Fourier transformation module with statistical averaging of the information signal (Fig. 4.10). The second one is based on a scalogram for displaying the wavelet transformation of the information signal (Fig. 4.11). It represents three axes: time (X axis), scale (Y axis), and coefficient value (Z axis), outlined with color.

Fig. 4.12 Current geomagnetic state in Solar Activity

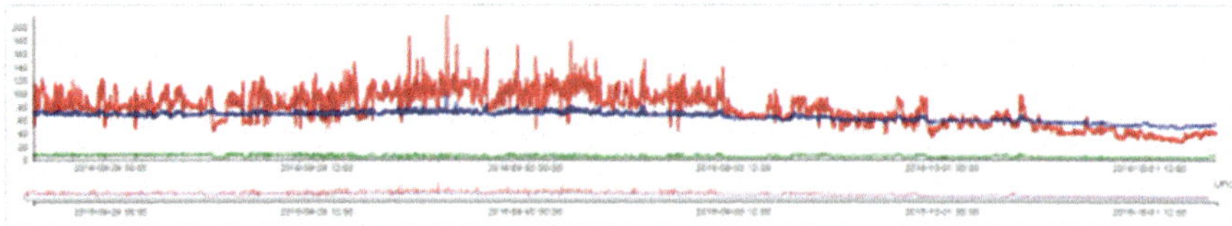

Fig. 4.13 Space weather data in the time domain in Solar Activity

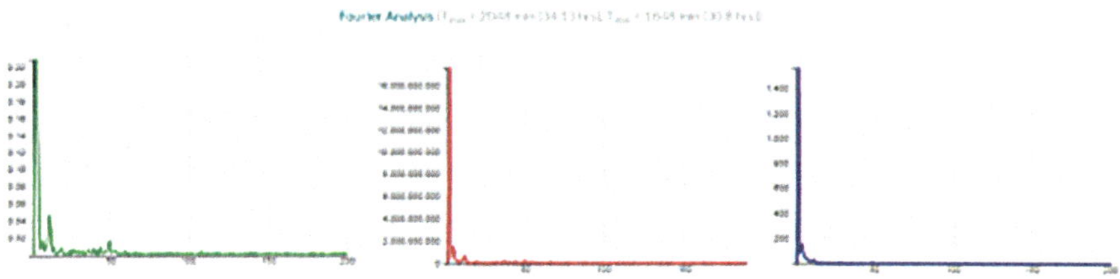

Fig. 4.14 Space weather data in the frequency domain in Solar Activity

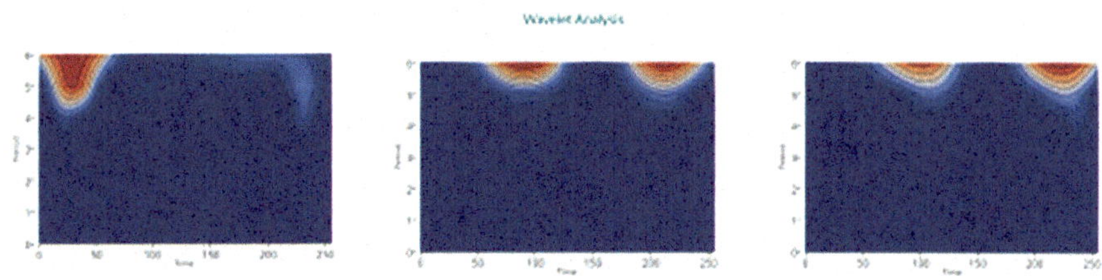

Fig. 4.15 Spectral analysis of space weather data in Solar Activity

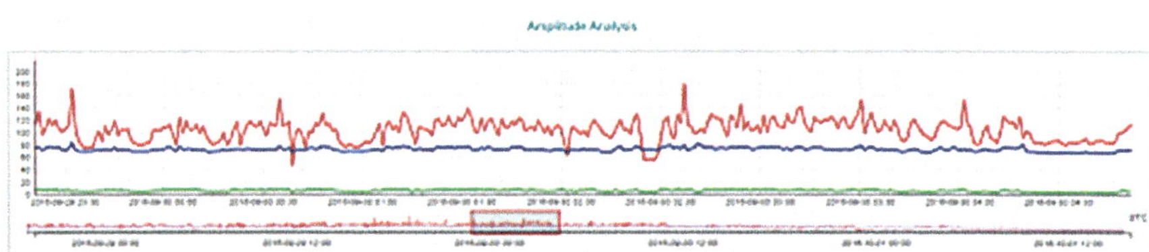

Fig. 4.16 Choosing the time period in Solar Activity

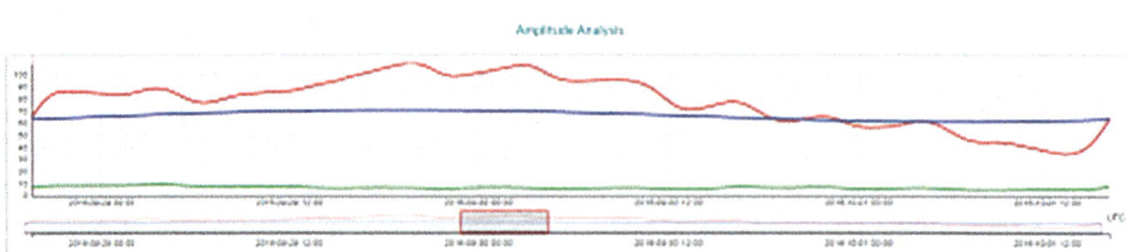

Fig. 4.17 Filtration of space weather data in Solar Activity

Fig. 4.18 Bandwidth value manual settings in Solar Activity

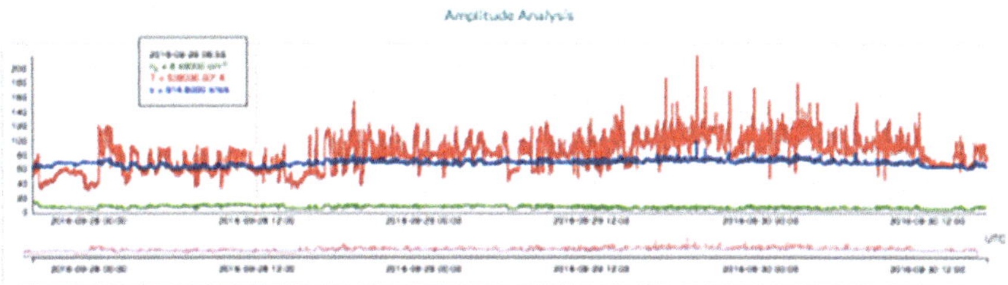

Fig. 4.19 Tooltips in Solar Activity

4.4.3 *The Magnetic Stations Service*

Nowadays, magnetic stations are connected to an integrated international network, which constitutes the main means of observing geomagnetic field. Within a period of a few hours, all data recorded by stations are sent to the world centers for collecting and processing the geomagnetic data:

- World Data Center for Solar–Terrestrial Physics, Moscow
- International Space Weather Center, regionally distributed in Brussels, Boulder, Moscow, Tokyo, Sydney, and Beijing
- Etc.

Geomagnetic data are represented in various formats, stored in a number of data sources, and it is hard to make simple queries to get them.

To aggregate them all together, the authors have included another service on the GEOMAGNET Web portal, which provides integrated access to all geomagnetic data from all available magnetic observatories. This service is called Magnetic Stations (https://www.geomagnet.ru/magnetic_stations.html).

The service has a single user interface, which provides a map displaying the locations of all available magnetic observatories. Due to this type of observatory representation, a user can choose any magnetic station in any location without the need to know its name or exact coordinates (Fig. 4.20). Also, a user can pick up an observatory and get its recorded data.

The service gets updated information about all active and inactive observatories. An active observatory is a magnetic station providing geomagnetic recorded data for the last 3 days. In turn, an inactive observatory is one that does not provide geomagnetic recorded data for the last 3 days. The list of active and inactive magnetic stations is updated on every loading of the service page, so a user can see actual information about the magnetic observatories efficiently and according to recorded (or missing) data.

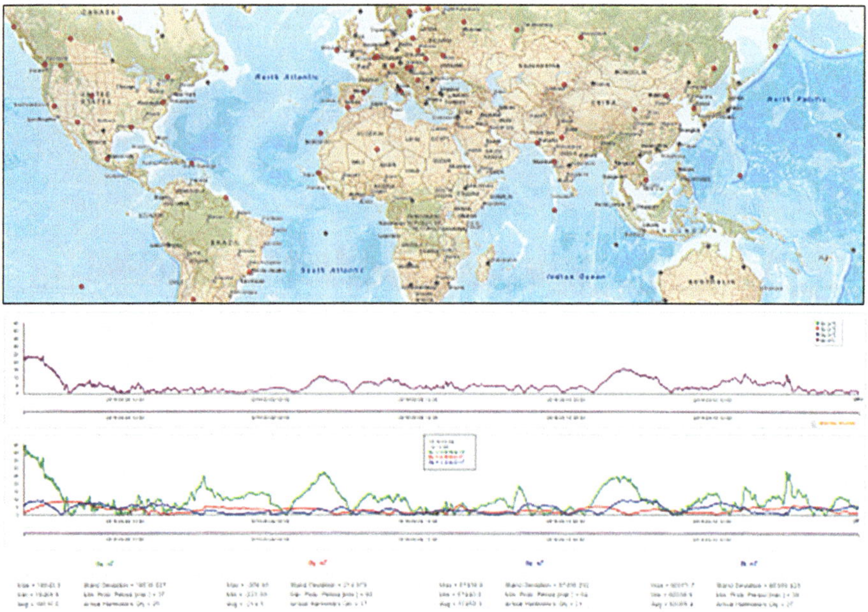

Fig. 4.20 Magnetic observatories on the map and their analysis results in magnetic stations

After page loading, a user has information about both active and inactive magnetic stations. All active observatories are colored red and inactive ones are colored gray to distinguish them from each other.

General information about the magnetic station (name, country, geographic coordinates, etc.) is available in pop-up windows, which are displayed over the observatory icon when a user moves a mouse pointer over it. This information is available for both active and inactive magnetic observatories.

More detailed information about a magnetic station supposes visualization and analysis of data recorded by the observatory for the last 3 days. The link to a separate page with data analysis for the observatory is available through the pop-up window with general information (the link is displayed only if the magnetic station is active; for inactive observatories only general information is available).

A page with the magnetic observatory data analysis loads independently from the main page of the Magnetic Stations service. So, a user can load a number of separate pages and compare data from various magnetic stations.

Data analysis in the Magnetic Stations service is based on the same principles as the analytical functionality of the Solar Activity service.

For every chosen magnetic observatory, the following parameters of geomagnetic field are visualized and analyzed within the service:

- North component (X) (nT)
- East component (Y) (nT)
- Down component (Z) (nT)
- Total intensity (nT)

In the time domain, these data are represented in two distinct charts (Fig. 4.21). One chart displays how the geomagnetic field north, east and down components vary over a 3-day time period. The other chart shows the time series of geomagnetic field total intensity values for the same period. Such separation gives a user the opportunity to view data in more details (showing all data on a single chart could cause some problems concerned with different orders of analyzing parameters).

Statistical data on the values recorded by every observatory are also available in the time domain, which contains maximal, minimal, and average values, and the variance and standard deviation, which are calculated for the north, east, and down components of the geomagnetic field as well as for its total intensity.

To reduce the noise in recorded values of geomagnetic field north, east, and down components and total intensity, the Magnetic Stations service provides linear and nonlinear filtration functions (Fig. 4.22). To switch on the filtration option, a user can choose "Filtered" mode in the "Service" menu. The optimal bandwidths for all geomagnetic field parameters are calculated automatically and are displayed on page loading. As in the Solar Activity service (see Sect. 4.4.2), the bandwidth can be selected by the user via special input fields corresponding to the analyzed parameters of geomagnetic field.

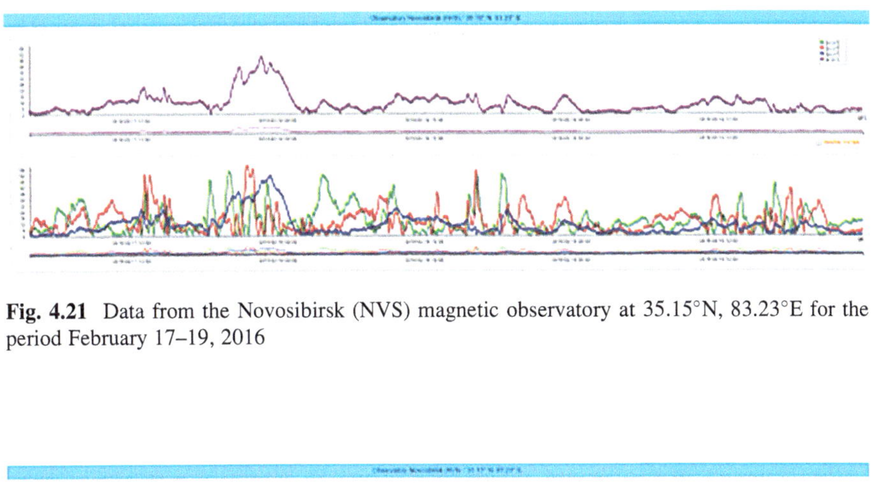

Fig. 4.21 Data from the Novosibirsk (NVS) magnetic observatory at 35.15°N, 83.23°E for the period February 17–19, 2016

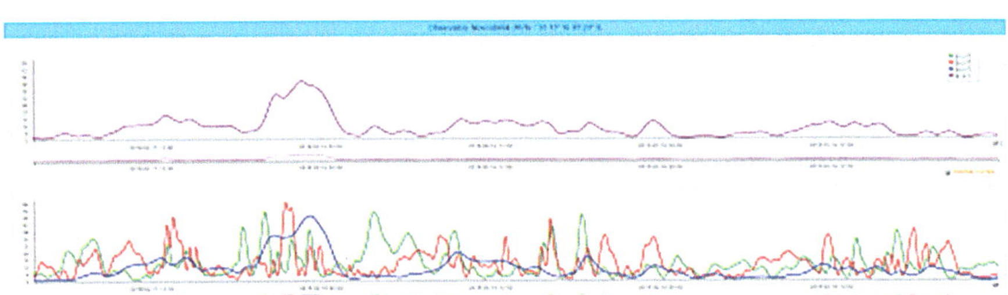

Fig. 4.22 Data from the Novosibirsk (NVS) magnetic observatory at 35.15°N, 83.23°E for the period February 17–19, 2016, after adaptive filtration

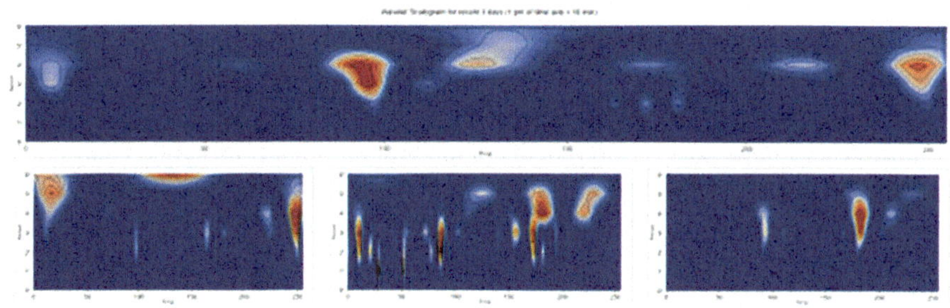

Fig. 4.23 Wavelet scalograms of data from the Cheongyang (CYG) magnetic observatory at 53.63°N, 126.854°E for the period February 19–21, 2016 (https://www.geomagnet.ru)

Spectral and time–frequency analysis in the Magnetic Stations service (as well as in Solar Activity service) is also represented in two charts. The first one is a periodogram, which is as an estimation of power spectral density based on calculation of square of data sequence Fourier transformation module with statistical averaging of the information signal. The second chart is an estimation of a wavelet scalogram of the information signal (Fig. 4.23).

4.5 Visualization of the Geomagnetic Field

It is widely accepted that data visualization is an important means of presenting abstract information from large volumes of raw data. This also applies to geomagnetic heterogeneous data. The best way to understand how the parameters of the geomagnetic field are distributed on the Earth's surface is to represent them graphically and visualize them. This means that abstract information becomes more understandable and provides insight into and knowledge of the geomagnetic data system in ways that were previously impossible.

In this section, an effective method for geomagnetic field parameter visualization is suggested, which is based on low-dimensional Web modeling (two and three dimensions).

The approach supposes data representation on two separate levels. The first one is the geographical description. It concerns the two- and three-dimensional visualization of the Earth's surface with variable zoom, along with detail parameters. Another level is an attributive one. This is a set of numerical values, which correspond to geomagnetic field parameter values for spatial coordinates with an appropriate step.

The basic type of representation of geospatial data in general and geomagnetic data in particular is the two-dimensional one. The approach realization requires at least two layers: the first layer is for the map visualization, and the second one is for the parameters of geomagnetic field representation.

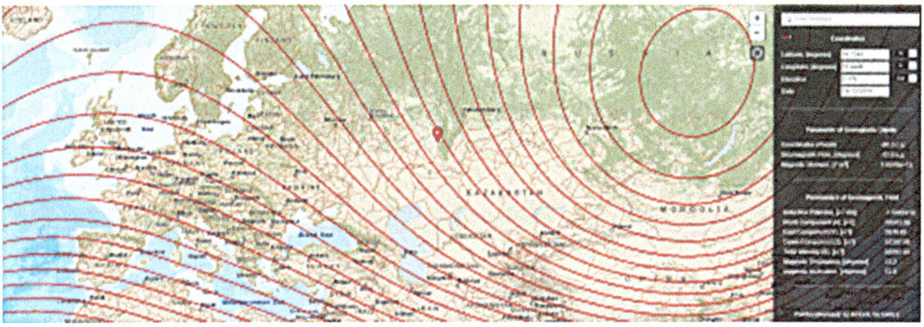

Fig. 4.24 Example of two-dimensional representation of the geomagnetic field in the GEOMAGNET WebGIS application

GIS-based realization of two-dimensional visualization is more often represented by using data streams in KML (or an archived version of it—KMZ) formats. Here KML stands for Keyhole Markup Language, which is an XML (i.e., platform-independent) markup language for describing geographical and attributive data (Wernecke 2008; Wilson 2008).

For example, Fig. 4.24 represents the functionality of the WebGIS Geomagnetic Calculator (WMM2015), which provides two-dimensional visualization of geomagnetic field parameters as a set of contour lines (curves connecting points of equal values of geomagnetic field parameters). By mouse clicking on the contour line, a user gets a tooltip window with information about the parameter and the value of the geomagnetic field there. This option is useful when it is necessary to study the global distribution of the Earth's magnetic field on the planet's surface. But this approach supposes that there is a single parameter to be analyzed and visualized. Actually, a lot of contour lines (even with different colors) could be visualized, but they would overlap each other, so the result efficiency of data graphical visualization would be lost.

The evolution of the two-dimensional approach supposes the addition of one more dimension. This is suitable for representing the parameters of the geomagnetic field where three-dimensional visualization techniques are used to integrate series of geomagnetic data into a form that is easier for the user to interpret.

Three-dimensional visualization of geomagnetic data also supposes two levels: the geographical and the attribute ones. Traditionally, such presentation of geographical data is a spherical Earth model (like a globe). It is the key to the whole visualization. All other attribute and geolocated data are placed over the "globe." Here, there are many representation types: gradients, histograms, and even flat contour lines (Butler 2006; Dalton 2013; Grossner and Goodchild 2008). All graphical representations are separated and can be studied independently as well as in connection with the other layers (Hu and Dai 2013; Huang et al. 2001).

Nowadays, the most popular technology for three-dimensional representation of geospatial data is based on Web components of a special type, which are known as virtual globes (Patterson 2007). Web applications using such technology are called geobrowsers. It is important to mention that geobrowser technology is based on the Earth's surface representation as a sphere with applied graphical layers.

Nowadays, one of the most popular versions of virtual globes is the Google Earth technology, which combines the possibilities of desktop and Web applications. This technology provides a wide variety of functions for visualization and multiaspect analysis of the Earth's surface (Blower 2010; Craglia et al. 2008; Craglia 2009; Henry 2009).

Figure 4.25 demonstrates an example of integrating the geomagnetic field parameter research results with the KML and Google Earth technologies. The geomagnetic field induction full vector is visualized as a set of contour lines. This approach provides an effective tool for complex analysis of various parameters of the geomagnetic field, correlations and principle definition. The numerical value of

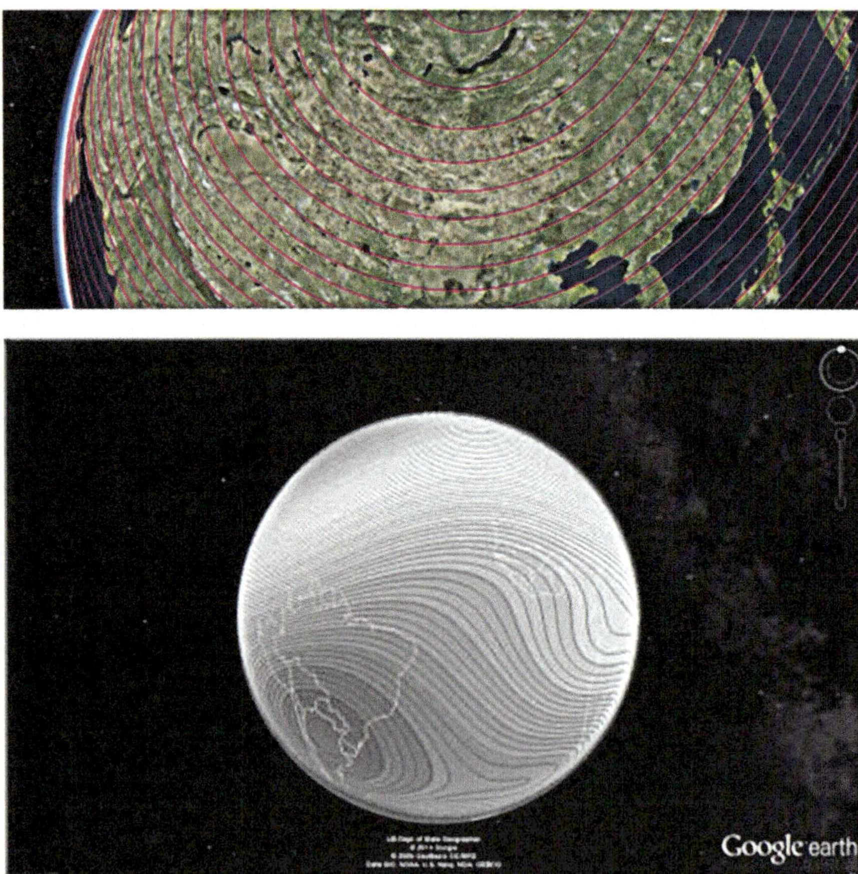

Fig. 4.25 Examples of three-dimensional representation of the geomagnetic field

the parameter (physical value) distributed along one contour line is by via picking the appropriate line using the mouse cursor.

On the GEOMAGNET Web portal, the three-dimensional representation of geomagnetic data is based on the technology of a software interface for accessing graphics hardware from within a Web browser without installing additional plug-ins.

A special API provides geobrowser integration with applications. This is a set of programming functions for creation, visualization, and many other types of manipulation of three-dimensional spatial data (Chow 2008; Coors 1998). The functionality of the API provides the Earth three-dimensional model inbuilt into the Web application and extends it with markers, info windows, analytical functions, etc.

The programming interface of geobrowser integration is used by an application as a set of local and/or remote functions. These functions are used with the special possibilities of an interpreter already used or additionally loaded on the user's computer.

In particular, in GEOMAGNET the realization of this functionality is based on the Cesium API—a set of modern technologies of globe rendering on the Web via HTML5, the most important of which is WebGL. The main advantage of Cesium is its complete integration into browsers, enabling a range of 3D graphics to be delivered on any device, anywhere. There is no plug-in needed (Fig. 4.26).

The Cesium API allows a developer to represent an interactive globe, which is responsive to any user actions (such as mouse clicks, scrolling up and down, etc.). This API also supports multilayer representation of the visualized parameters.

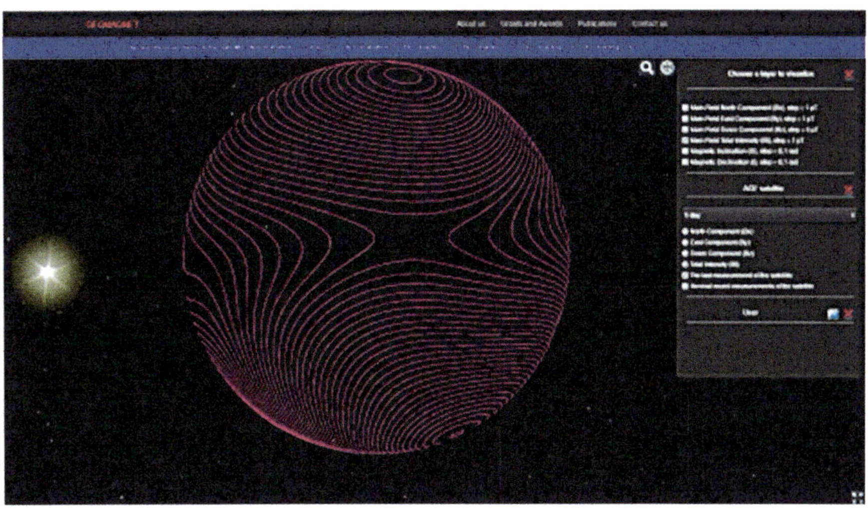

Fig. 4.26 Service of three-dimensional representation of the geomagnetic field in the GEOMAGNET WebGIS application

On the GEOMAGNET Web portal, the three-dimensional representation divides a window into two parts: one for the geomagnetic data in a three-dimensional representation; and another one for manipulating the view (choosing and combining layers to be displayed, zooming, changing the observation point, etc.).

Within the three-dimensional representation, visualization on the virtual globe of either a single parameter or multiple parameters of the geomagnetic field includes, among others:

- Main field north component
- Main field east component
- Main field down component
- Main field total intensity
- Magnetic inclination
- Magnetic declination

The resolution of geomagnetic data visualization by default is set to 1 μT for the main field north and east components, and total intensity, 5 μT for the main field down component, and 0.1 rad for magnetic inclination and declination. In order to increase or decrease the given resolution, a user can manually change these values.

The default settings suppose the data representation as a set of contour lines, which are distributed on the globe surface. However, it is possible to display more than one parameter at the same time in one window. The series of parameters for more understandable presentation are visualized as a set of histograms with different colors. With this option a user can compare multiple parameters on one screen just by combining the displayed layers.

Three-dimensional representation on the GEOMAGNET Web portal also supposes an option to switch the view to the two-dimensional view or Columbus view (2.5D) and back. This opportunity is provided by the Cesium API and is available via a globe icon in the top left corner of the window (Fig. 4.26).

The Cesium API functionality provides a zooming option to view some chosen regions on the globe with more details. The zoom tool is also located in the top left corner of the window (Fig. 4.26) and is available via a user mouse click (also, there is another way to turn on zooming: a user can scroll the mouse wheel up and down).

4.6 Conclusion

Dedicated to GIS/WebGIS approaches in geomagnetic field monitoring, this chapter, among other things, has presented:

- Analysis of modern tendencies of GIS evolution along with the future direction for the development, realization, and implementation of a special GIS based on modern Web technologies, providing great functionality by integrating calculation of parameters of geomagnetics, its variations and its anomalies, and visualizing the calculated distribution in terrestrial and circumterrestrial space.

- During their research, the authors have suggested and developed a conception for modeling and distant analytical description of space weather, the geomagnetic field, and its variation parameters. The conception is based on a paradigm of combination of heterogeneous data sources (satellite, air, marine, and ground observations, and magnetic observatories) into an integrated information space with platform-independent mechanisms for DSP and data interpretation. The research proved that a program formalizing the conception of the GEOMAGNET Web portal (https://www.geomagnet.ru) reduced the time required for collecting, processing, and analyzing heterogeneous data about space weather and the geomagnetic field and its variations. The GEOMAGNET Web portal provides a complex calculation, analysis, and 2D/3D visualization of the geomagnetic field and its variation parameters. The geomagnetic field and variations models represented and described by GEOMAGNET meet the requirements of specialists in various areas. They effectively provide formatting and structuring of calculated data on the Earth's magnetosphere parameters and their analysis. An innovative three-dimensional dynamic model of geomagnetic field parameter distribution in circumterrestrial space, based on Web technologies and technologies of modern geobrowsers, is presented. This model provides a multilayer visualization as an effective tool for complex description and analysis of various geophysical parameters and definition of the appropriate correlations and principles.

References

Alekseev ER, Chesnokov OV, Rudchenko EA (2008) Scilab. Solving mathematical and engineering problems. Binom, Moscow (in Russian)

Alesheikh AA, Helali H, Behroz HA (2002) Web GIS: technologies and its applications. ISPRS technical commission IV symposium 2002, Ottawa, Canada. pp 1–9

Amazon Web Services (2012) Mapping and geospatial analysis in Amazon web services using ArcGIS. http://media.amazonwebservices.com/AWS_ESRI_Mapping_GeoSpatial_Analysis_Using_ArcGIS.pdf

Blower JD (2010) GIS in the cloud: implementing a web map service on Google App engine. In: Proceedings of the 1st international conference and exhibition on computing for geospatial research, New York, NY, USA

Bogoutdinov SR, Gvishiani AD, Agayan SM, Solovyev AA, Kihn E (2010) Recognition of disturbances with specified morphology in time series. Part 1: Spikes on magnetograms of the worldwide INTERMAGNET network. Izv Phys Solid Earth 46(11):1004–1016

Butler D (2006) Virtual globes: the web-wide world. Nature 439:776–778

Campbell WH (2003) Introduction to geomagnetic fields, 2nd edn. Cambridge University Press, Cambridge

Carver S (2001) Public participation using Web-based GIS. Environ Plann B Plann Des 28:803–804

Chow T (2008) The potential of maps APIs for Internet GIS applications. Trans GIS 12:179–191

Coors V (1998) Extended abstract on International workshop for interactive applications of mobile computing "Using Wearable GIS in outdoor applications", Germany

Craglia M (2009) From INSPIRE to digital Earth: research challenges for the next generation spatial data infrastructures. In: Proceedings of the 12th AGILE conference, Hanover, Germany
Craglia M, Goodchild MF, Annoni A, Camara G, Gould M, Kuhn M, Mark D, Masser I, Maguire D, Liang S, Parsons E (2008) Next generation digital earth. Inter J Spatial Data Infrastruct Res 3:146–167
Dalton CM (2013) Sovereigns, spooks, and hackers: an early history of Google Geo Services and map mashups. Cartographica 48:261–274
De Smith MJ, Longley PA, Goodchild MF (2007) Geospatial analysis: a comprehensive guide to principles, techniques and software tools. Winchelsea Press, Winchelsea
ESRI Inc. (1999) MapObjects Internet map server user guide, ESRI, Inc., California, USA
Flanagan D (2011) JavaScript: the definitive guide, 6th edn. O'Reilly Media Inc., Sebastopol
Goodchild MF (2008) The use cases of digital earth. Int J Dig Earth 1:31–42
Goodchild MF, Fu P, Rich P (2007) Sharing geographic information: an assessment of the geospatial one-stop. Ann Assoc Am Geogr 97(2):250–266
Gore A (1999) The digital earth: understanding our planet in the 21st century. Photogramm Eng Remote Sens 65:528
Green D, Bossomaier T (2002) Online GIS and spatial metadata. Taylor & Francis, London
Grossner K, Goodchild M (2008) Defining a digital earth system. Trans GIS 12:145–160
Gupta A, Sharma MP (2014) Development of Web GIS based digital avalanche atlas. Int J Sci Eng Technol Res 3(4):117–124
Haklay M, Singleton A, Parker C (2008a) Web mapping 2.0: the neogeography of the geo web. Geogr Compass 6(2):2011–2039
Hall B, Leahy MG (2008) Open source approaches in spatial data handling, Advances in geographic information science. Springer, Berlin
Harvey MJ, Han J (2009) Geographic data mining and knowledge 43 discovery. Chapman & Hall, Boca Raton
Henry A (2009) Supporting document: using Google Earth for light-weight Internet GIS, MSc dissertation, University of Edinburgh
Hu SF, Dai T (2013) Online map application development using Google maps API, SQL database, and ASP.NET. Int J Inf Commun Techno Res 3(3):102–111
Huang B, Jiang B, Li H (2001) An integration of GIS, virtual reality and the Internet for visualization, analysis and exploration of spatial data. Int J Geogr Inf Sci 15:439–456
Kim D, Kim M (2002) Web GIS service component based on open environment. In: Proceedings of IGARSS geoscience and remote sensing symposium. IEEE Press, June 2002, pp 3346–3348
Longley A, Goodchild F, Maguire J, Rhind J (2001) Geographical information systems and science, 2nd edn. Wiley, Chichester
Mandea M, Korte M (eds) (2011) Geomagnetic observations and models, IAGA special sopron series vol 5. Springer, New York
Mapes J (2008) Mapping for the masses: Google mapping tools. Vis Stud 23:78–84
McLaughlin J, Groot R (2000) Geospatial data infrastructure: concepts, cases and good practice, Spatial information systems and geostatistics series. Oxford University Press, Oxford
Merrill RT, McElhinny MW, McFadden PL (1996) The magnetic field of the Earth. Academic, San Diego
Patterson TC (2007) Google earth as a (not just) geography education tool. J Geogr 106:145–152
Peng Z, Tsou MH (2003) Internet GIS: distributed geographic information services for the Internet and wireless networks. Wiley, Hoboken, p 720
Samardak AS (2005) Geographic information systems. Far Eastern State University Publishing House, Vladivostok (in Russian)
Soloviev A, Bogoutdinov S, Gvishiani A, Kulchinskiy R, Zlotnicki J (2013) Mathematical tools for geomagnetic data monitoring and the INTERMAGNET Russian segment. Data Sci J 12:114–119
Svennerberg G (2010) Beginning Google maps API 3. Apress & Springer, New York
Tsvetkov A (1998) Geo information systems and technologies. Moscow (in Russian)

Vorobev AV, Shakirova GR (2015a) Application of geobrowsers to 2D/3D–visualization of geomagnetic field. In: 15th international multidisciplinary scientific Geo conference and EXPO, SGEM 2015, vol I

Vorobev AV, Shakirova GR (2015b) Modeling and 2D/3D–visualization of geomagnetic field and its variations parameters. In: Proceedings of GISTAM 2015—1st international conference on geographical information systems theory, applications and management, 2015

Vorobev AV, Shakirova GR (2015c) Web-based information system for modeling and analysis of parameters of geomagnetic field. Proc Comput Sci 59:73–82. ELSEVIER

Vorobev AV, Shakirova GR et al (2014) Questions of construction of geographic information systems on the basis of well-known web mapping services. Acad J West Sib 10(2):10 (in Russian)

Wernecke J (2008) The KML handbook. Addison-Wesley, Boston

Wilson T (2008) OGC KML 2.2.0, OGC

Yang C, Wu H, Huang Q, Li Z, Li J (2011) Using spatial principles to optimize distributed computing for enabling the physical science discoveries. Proc Natl Acad Sci 108 (14):5498–5503

Zhurkin IG, Shaytura SV (2009) Geographic information systems. Kudits-Press, Moscow. p 272 (in Russian)

Appendix

Listing Part of *GEOmagnetic_v*1.0 Programming Code

```
function [U, Bx, By, Bz, B, D, I]= GMF(north, east, alt, year)

R=6371.032;
a=6378.13649;
b=6356.75175;
Age=2010.0
IGRF_g=[ 0 0 0 0 0 0 0 0 0 0 0 0 0 0;
-29496.5 -1585.9 0 0 0 0 0 0 0 0 0 0 0;
-2396.6 3026.0 1668.6 0 0 0 0 0 0 0 0 0;
1339.7 -2326.3 1231.7 634.2 0 0 0 0 0 0 0 0 0;
912.6 809.0 166.6 -357.1 89.7 0 0 0 0 0 0 0 0;
-231.1 357.2 200.3 -141.2 -163.1 -7.7 0 0 0 0 0 0 0;
72.8 68.6 76.0 -141.4 -22.9 13.1 -77.9 0 0 0 0 0 0;
80.4 -75.0 -4.7 45.3 14.0 10.4 1.6 4.9 0 0 0 0 0;
24.3 8.2 -14.5 -5.7 -19.3 11.6 10.9 -14.1 -3.7 0 0 0 0;
5.4 9.4 3.4 -5.3 3.1 -12.4 -0.8 8.4 -8.4 -10.1 0 0 0;
-2.0 -6.3 0.9 -1.1 -0.2 2.5 -0.3 2.2 3.1 -1.0 -2.8 0 0;
3.0 -1.5 -2.1 1.6 -0.5 0.5 -0.8 0.4 1.8 0.2 0.8 3.8 0;
-2.1 -0.2 0.3 1.0 -0.7 0.9 -0.1 0.5 -0.4 -0.4 0.2 -0.8 0];

IGRF_h= [0 0 0 0 0 0 0 0 0 0 0 0 0 0;
0 4945.1 0 0 0 0 0 0 0 0 0 0 0;
0 -2707.7 -575.4 0 0 0 0 0 0 0 0 0;
0 -160.5 251.7 -536.8 0 0 0 0 0 0 0 0 0;
0 286.4 -211.2 164.4 -309.2 0 0 0 0 0 0 0 0;
0 44.7 188.9 -118.1 0.1 100.9 0 0 0 0 0 0 0;
0 -20.8 44.2 61.5 -66.3 3.1 54.9 0 0 0 0 0 0;
```


S. Kolios et al., *GIS and Environmental Monitoring*, Geotechnologies and the Environment 20, DOI 10.1007/978-3-319-53086-4

```
0 -57.8 -21.2 6.6 24.9 7.0 -27.7 -3.4 0 0 0 0 0;
0 10.9 -20.0 11.9 -17.4 16.7 7.1 -10.8 1.7 0 0 0 0;
0 -20.5 11.6 12.8 -7.2 -7.4 8.0 2.2 -6.1 7.0 0 0 0;
0 2.8 -0.1 4.7 4.4 -7.2 -1.0 -4.0 -2.0 -2.0 -8.3 0 0;
0 0.1 1.7 -0.6 -1.8 0.9 -0.4 -2.5 -1.3 -2.1 -1.9 -1.8 0;
0 -0.8 0.3 2.2 -2.5 0.5 0.6 0.0 0.1 0.3 -0.9 -0.2 0.8]

SV_g=[0 0 0 0 0 0 0 0 0 0 0 0 0 0 0;
11.4 16.7 0 0 0 0 0 0 0 0 0 0 0;
-11.3 -3.9 2.7 0 0 0 0 0 0 0 0 0 0;
1.3 -3.9 -2.9 -8.1 0 0 0 0 0 0 0 0 0;
-1.4 2.0 -8.9 4.4 -2.3 0 0 0 0 0 0 0 0;
-0.5 0.5 -1.5 -0.7 1.3 1.4 0 0 0 0 0 0 0;
-0.3 -0.3 -0.3 1.9 -1.6 -0.2 1.8 0 0 0 0 0 0;
0.2 -0.1 -0.6 1.4 0.3 0.1 -0.8 0.4 0 0 0 0 0;
-0.1 0.1 -0.5 0.3 -0.3 0.3 0.2 -0.5 0.2 0 0 0 0;
0 0 0 0 0 0 0 0 0 0 0 0 0;
0 0 0 0 0 0 0 0 0 0 0 0 0;
0 0 0 0 0 0 0 0 0 0 0 0 0;
0 0 0 0 0 0 0 0 0 0 0 00]

SV_h=[0 0 0 0 0 0 0 0 0 0 0 0 0 0 0;
0 -28.8 0 0 0 0 0 0 0 0 0 0 0;
0 -23.0 -12.9 0 0 0 0 0 0 0 0 0 0;
0 8.6 -2.9 -2.1 0 0 0 0 0 0 0 0 0;
0 0.4 3.2 3.6 -0.8 0 0 0 0 0 0 0 0;
0 0.5 1.5 0.9 3.7 -0.6 0 0 0 0 0 0 0;
0 -0.1 -2.1 -0.4 -0.5 0.8 0.5 0 0 0 0 0 0;
0 0.6 0.3 -0.2 -0.1 -0.8 -0.3 0.2 0 0 0 0 0;
0 0.0 0.2 0.5 0.4 0.1 -0.1 0.4 0.4 0 0 0 0;
0 0 0 0 0 0 0 0 0 0 0 0 0;
0 0 0 0 0 0 0 0 0 0 0 0 0;
0 0 0 0 0 0 0 0 0 0 0 0 0;
0 0 0 0 0 0 0 0 0 0 0 0 0]

lamda=east*(%pi/180);

fi=north*(%pi/180);
fi_sh=atan(((b^2+alt*sqrt(a^2*(cos(fi))^2+b^2*(sin(fi))^2))/
(a^2+alt*sqrt(a^2*(cos(fi))^2+b^2*(sin(fi))^2)))*tan(fi))
tetta=(%pi/2)-fi_sh
r=sqrt(alt^2+2*alt*sqrt(a^2*(cos(fi))^2+b^2*(sin(fi))^2)+((a^4*
(cos(fi))^2+b^4*(sin(fi))^2)/(a^2*(cos(fi))^2+b^2*(sin(fi))^2)))
F_IGRF_g=IGRF_g+(SV_g)*(year-Age);
F_IGRF_h=IGRF_h+(SV_h)*(year-Age);
```

```
IGRF_N=size(F_IGRF_g, 'r');
N=IGRF_N-1;
function [rez]=PI_odd(i)
rez=1;
   for k=1:i
rez1=2*k-1;
rez=rez*rez1;
end
funcprot(0)
endfunction
function [eps]=epsilon(i)
    if i<1 then eps=1;
        else eps=2;
    end
funcprot(0)
endfunction
_U(_r,_lamda,_tetta)****************
function [potential]=_U(_r, _lamda, _tetta)
sum_n=0
for n=1:N
sum_m=0
             for m=0:n
half_REZ=R*(F_IGRF_g(n+1,m+1)*cos(m*_lamda)+  F_IGRF_h(n+1,m+1)
*sin(m*_lamda))*(R/_r)^(n+1)*..
PI_odd(n)*sqrt(epsilon(m)/(factorial(n+m)*factorial(n-m)))*(sin
(tetta))^m*..
((cos(_tetta))^(n-m)-..
((n-m)*(n-m-1)/(2*(2*n-1)))*(cos(_tetta))^(n-m-2)+..
((n-m)*(n-m-1)*(n-m-2)*(n-m-3)/(2*4*(2*n-1)*(2*n-3)))*(cos(_tet-
ta))^(n-m-4)-..
((n-m)*(n-m-1)*(n-m-2)*(n-m-3)*(n-m-4)*(n-m-5)/(2*4*6*(2*n-1)*
(2*n-3)*(2*n-5)))*(cos(_tetta))^(n-m-6)+..
((n-m)*(n-m-1)*(n-m-2)*(n-m-3)*(n-m-4)*(n-m-5)*(n-m-6)*(n-m-7)/
(2*4*6*8*(2*n-1)*(2*n-3)*(2*n-5)*(2*n-7)))*(cos(_tetta))^(n-m-
8)-..
((n-m)*(n-m-1)*(n-m-2)*(n-m-3)*(n-m-4)*(n-m-5)*(n-m-6)*(n-m-7)*
(n-m-8)*(n-m-9)/(2*4*6*8*10*(2*n-1)*(2*n-3)*(2*n-5)*(2*n-7)*
(2*n-9)))*(cos(_tetta))^(n-m-10)+..
((n-m)*(n-m-1)*(n-m-2)*(n-m-3)*(n-m-4)*(n-m-5)*(n-m-6)*(n-m-7)*
(n-m-8)*(n-m-9)*(n-m-10)*(n-m-11)/(2*4*6*8*10*12*(2*n-1)*(2*n-3)
*(2*n-5)*(2*n-7)*(2*n-9)*(2*n-11)))*..
(cos(_tetta))^(n-m-12));

sum_m=sum_m+half_REZ;
end
```

```
sum_n=sum_n+sum_m;
    end
  potential=sum_n;
funcprot(0)
endfunction

function f=dUdtetta(_tetta)
  f=_U(r,lamda,_tetta)
endfunction
BX = (1/r)*numderivative(dUdtetta,tetta);

function f=dUdlamda(_lamda)
  f=_U(r,_lamda,tetta)
endfunction
BY = (-1/(r*sin(tetta)))*numderivative(dUdlamda,lamda);

function f=dUdr(_r)
  f=_U(_r,lamda,tetta)
endfunction
BZ =numderivative(dUdr,r)
U=_U(r,lamda,tetta)
Bx = BX*cos(fi-fi_sh)+BZ*sin(fi-fi_sh)
By = BY
Bz = BZ*cos(fi-fi_sh)-BX*sin(fi-fi_sh)
B = sqrt((Bx^2)+(By^2)+(Bz^2))
D = atan(By/Bx)
I = asin(Bz/B)
funcprot(0)

endfunction

printf("LAT, LONG, U, Bx, By, Bz, B, D, I")
printf("\n");
NorthArray = [-41:-1:-80];
EastArray = [-180:1:179];
AltArray = 0;
YearArray = 2010.0;

for i=1:length (NorthArray)
for j=1:length (EastArray)

[U,Bx,By,Bz,B,D,I]=
GMF(NorthArray(i),EastArray(j),AltArray,YearArray)
u=U
bx=Bx
```

```
by=By
bz=Bz
b=B
dd=D
ii=I

u(i,j)=U
bx(i,j)=Bx
by(i,j)=By
bz(i,j)=Bz
b(i,j)=B
dd(i,j)=D
ii(i,j)=I

printf("%3.1f",NorthArray(i));printf(",")
printf("%3.1f",EastArray(j));printf(",")
printf("%3.0f",u(i,j));printf(",")
printf("%3.0f",bx(i,j));printf(",")
printf("%3.0f",by(i,j));printf(",")
printf("%3.0f",bz(i,j));printf(",")
printf("%3.0f",b(i,j));printf(",")
printf("%3.5f",dd(i,j));printf(",")
printf("%3.5f",ii(i,j));printf(",")
printf("\n");

end
end
```

Glossary

Application Program Interface (API) A set of routines, protocols, and tools for building software applications.

Attribute Descriptive characteristic of a feature. An attribute can be value of a measurement corresponding to a feature.

Base image/map An image (or a map) portraying background geographic information on which other layers of information can be placed. Base images/maps display natural Earth surface features and permanent spatial objects.

Boolean algebra Mathematical field using logical (Boolean) operators like *AND*, *OR*, and *NOT* to represent/combine/exempt values and information.

Buffer A polygon enclosing an area within a specified distance from a point, line, or other polygon.

Classification The organization of data into categories using a set of criteria and algorithms.

Contour The outline of an object or isopleths in a spatial data file (for example, isoheights, isotherms, etc.).

Control points Points on the surface of the Earth of known coordinates, used to georeference data.

CodeIgniter An open-source Model–View–Controller (MVC) framework written in PHP.

Coordinate reference system A projection reference system used to define the position of a point on the Earth's surface and locate geographical entities. A coordinate reference system can be local, regional, or global.

Cloud GIS The integration of GIS software and its related services into cloud services providing more GIS capabilities through the Web. The cloud GIS applications operate in a shared data center that is accessed via the Internet.

Data as a Service (DaaS) One of the four basic service models in the cloud computing environment. It is based on the concept that the data are provided on demand to the user regardless of the geographic or organizational separation of the provider and consumer.

S. Kolios et al., *GIS and Environmental Monitoring*, Geotechnologies and the Environment 20, DOI 10.1007/978-3-319-53086-4

Digital elevation model (DEM) A data file with terrain elevation values. A DEM can be represented as a raster (a grid of squares, also known as a height-map when presenting elevation) or as a vector-based triangular irregular network (TIN) file.

Digitization The process of converting an analog image or map into a digital format.

Entities Items presenting the stored information. Items may be tangible (or not) and further defined by their attributes.

Exospace magnetic activity The complex geomagnetic disturbances caused by the increased magnetic activity of celestial bodies (except the Sun) from both the solar system (Jupiter, Saturn, Uranus, Neptune, comets, etc.) and outer space.

Feature A set of data with common attributes (for example, an area that represents a lake and is defined by a polygon). A feature encompasses both entity and object in the spatial dataset.

Format Specific structure of characters and values in order to store data measurements, estimates, and characteristics, as well as metadata information.

Geographical Information System (GIS) An integrated system (consisting of software and hardware) designed to create/capture, store, analyze, manage, and visualize any type of spatial (and/or nonspatial) data and information.

Georeference The procedure to provide geographic dimensions as a digital image (or data file), which contains spatial information.

Geomagnetic anomalies Intraterrestrial natural sources presenting distortion of magnetic intensity vectors. They are caused by the magnetization of the crust top parts.

Geomagnetic field The Earth's magnetic field extending from the Earth's interior to where it meets the solar wind (a stream of charged particles emanating from the Sun).

Grid A network of horizontal and vertical lines designed to better pinpoint any location on the Earth by laying a vertical and horizontal grid over the Earth's layout. The vertical lines are called the *longitude* and the horizontal lines are called the *latitude*. The intersection of these two lines determines exact locations on the Earth surface.

Infrastructure as a Service (IaaS) One of the four basic service models in the cloud computing environment. It is a standardized, highly automated, cloud computing model in which a third-party provider hosts hardware, software, servers, storage, and other infrastructure components on behalf of its users (computing resources complemented by storage and networking capabilities are owned and hosted by the service provider and offered to customers on demand).

Interactive (map) Digital illustration of an area having geographic dimensions and spatial characteristics, allowing two-way electronic communication (interaction) between the user and the digital file (digital map).

Interpolation (spatial) The procedure to estimate continuous predicted surface values based on point measurements; this type of interpolated surface is often called a *statistical surface*. Alternatively, spatial interpolation is the process of using points with known values to estimate values at other, unknown points.

Layer The visual representation of a spatial georeferenced dataset at the graphical user interface in the GIS, which could correspond to a legend item on a paper map.

Map tiles Parts (image parts) of a digital interactive map displayed in a browser.

Magnetosphere The solar wind interacts with this barrier, flows around the Earth, and creates a special region with the geomagnetic field inside. This region is known as the Earth's magnetosphere.

Map Algebra An analysis language used for performing cartographic spatial analysis in raster data. Map Algebra uses math-like expressions containing operators and functions.

Model-view-controller (MVC) A software architectural methodology for designing and implementing user interfaces. It divides any software application into three structural modules, so as to separate the internal handling of information from the way that the information is presented to or accepted by the user.

Pixel Shortcut for *pixel element*, i.e., the smallest discrete elements that compose an image.

Plasma layer The region filled with low-energy particles (having values at fractions of kiloelectron volts) with rather a high concentration. It is characterized by its anisotropic distribution by pitch angles with a predominance of particles with longitudinal speeds.

Platform as a Service (PaaS) One of the four basic service models in the cloud computing environment. It is a cloud computing model that delivers applications over the Internet. It provides platform services on which developers can build and deploy custom applications.

Query A set of conditions (*questions*) used to retrieve specific information from a database.

Raster A general type of data file where for every pixel a specific value is stored. The total number of pixels comprise a gridded pattern comprising a digital image.

Remote sensing The science, technology, and art of obtaining information about an object, area, or phenomenon by analyzing data acquired by a device that is not in physical direct contact with the object, area, or phenomenon under investigation.

Software as a Service (SaaS) One of the four basic service models in the cloud computing environment. It is a software licensing and delivery model in which software is licensed on a subscription basis and is centrally hosted.

Slope A slope is defined by a plane tangent to a topographic surface, as modeled by the digital elevation model (DEM) at a point.

Spatial reference system A set of mathematical transformations for spherical geometry in order to provide geographical information and to project points/areas on the real Earth's curve. The spatial reference system includes different choices of geographic coordinate systems and map projections according to the needs for geographic accuracy on local, national, and global scales.

Solar wind A flow of ionized particles (they are mostly helium–hydrogen plasma), which radially flow from the solar corona to outer space. This natural phenomenon causes *magnetic storms*.

Schumann resonance The phenomenon of formatting standing electromagnetic waves with low, superlow, and extremely low frequencies between the Earth's surface and the ionosphere.

Vector A general type of data file, where geometrical objects such as points, lines, curves, and shapes or polygons—all of which are based on mathematical expressions and geographical entities—are presented as separate objects.

Web coverage service (WCS) The Web-based retrieval of coverages for digital geospatial information. They represent space/time–varying phenomena useful for client-side rendering, as input into existing models.

Web Feature Service (WFS) This allows a client to retrieve and update geospatial data encoded in Geography Markup Language (GML) format. The Web Feature Services (WFS) standard specifies a number of different request types.

WebGIS Any GIS available through Web technologies that allows all end users to simultaneously view the same up-to-date spatial data through a Web-based interface.

Web Map Services (WMS) A standard protocol for serving georeferenced map images generated by a map server using data from a GIS database. The WMS protocol offers a standardized method to access maps on various servers. The application through the WMS servers can collect different sources of information, reproject the map layers (if needed), and send them back as a combined image containing all requested data layers. For example, one server may offer a topographic base map and other servers may offer thematic layers.

Index

S. Kolios et al., *GIS and Environmental Monitoring*, Geotechnologies and the Environment 20, DOI 10.1007/978-3-319-53086-4

Printed by Printforce, the Netherlands